COMPUTER SCIENCE, TECHNOLOGY AND APPLICATIONS

CREATING DIGITAL COMMUNITIES

A RESOURCE TO DIGITAL INCLUSION

Computer Science, Technology and Applications

Additional books in this series can be found on Nova's website under the Series tab.

Additional e-books in this series can be found on Nova's website under the e-book tab.

COMPUTER SCIENCE, TECHNOLOGY AND APPLICATIONS

CREATING DIGITAL COMMUNITIES

A RESOURCE TO DIGITAL INCLUSION

SLOANE CARUSO
EDITOR

nova publishers
New York

For permission to use material from this book please contact us:
Telephone 631-231-7269; Fax 631-231-8175
Web Site: http://www.novapublishers.com

Library of Congress Cataloging-in-Publication Data

ISBN: 978-1-63117-631-9

Published by Nova Science Publishers, Inc. † New York

CONTENTS

PREFACE

Digital inclusion is the ability of individuals and groups to access and use information and communication technologies. Not all members of a community benefit equally, and some communities have been left out altogether. This book is a resource for fostering digital inclusion throughout the United States so that everyone can take advantage of digital technologies. The framework is structured around a vision for the future, principles that define digital inclusion, goals to make digital inclusion a reality, and sample strategies for achieving these goals.

Chapter 1 – Digital inclusion is the ability of individuals and groups to access and use information and communication technologies. Not all members of a community benefit equally, and some communities have been left out altogether. Building Digital Communities: A framework for action is a resource for fostering digital inclusion throughout the United States so that everyone can take advantage of digital technologies. The Framework is structured around a vision for the future, principles that define digital inclusion, goals to make digital inclusion a reality, and sample strategies for achieving the goals.

Chapter 2 – The "digital divide" is a term that has been used to characterize a gap between "information haves and have-nots," or in other words, between those Americans who use or have access to telecommunications and information technologies and those who do not. One important subset of the digital divide debate concerns high-speed Internet access and advanced telecommunications services, also known as broadband. Broadband is provided by a series of technologies (e.g., cable, telephone wire, fiber, satellite, wireless) that give users the ability to send and receive data at volumes and speeds far greater than traditional "dial-up" Internet access over telephone lines. Broadband technologies are currently being deployed primarily by the private sector throughout the United States. While the numbers of new broadband subscribers continue to grow, studies and data suggest that the rate of broadband deployment in urban/suburban and high income areas is outpacing deployment in rural and low-income areas. Some policymakers, believing that disparities in broadband access across American society could have adverse economic and social consequences on those left behind, assert that the federal government should play a more active role to avoid a "digital divide" in broadband access. With the conclusion of the grant and loan awards established by the American Recovery and Reinvestment Act of 2009 (P.L. 111-5), there remain two ongoing federal vehicles which direct federal money to fund broadband infrastructure: the broadband and telecommunications programs at the Rural Utilities Service (RUS) of the U.S. Department of Agriculture and the Universal Service Fund (USF) programs under the Federal Communications Commission (FCC). Although the USF's High Cost Program does not explicitly fund broadband infrastructure, subsidies are used, in many cases, to upgrade existing telephone networks so that they are capable of delivering high-speed services. Additionally, subsidies provided by USF's Schools and Libraries Program and Rural Health Care Program are used for a variety of telecommunications services, including broadband access. Currently the USF is undergoing a major transition to the Connect America Fund, which is targeted to the deployment, adoption, and utilization of both fixed and mobile broadband. To the extent that the 113th Congress may consider various options for further encouraging broadband deployment and adoption, a key issue is how to strike a balance between providing federal assistance for unserved and underserved areas where the private sector may not be providing acceptable levels of broadband service, while at the same time minimizing any deleterious effects that government intervention in the marketplace may have on competition and private sector investment.

In: Creating Digital Communities
Editor: Sloane Caruso
ISBN: 978-1-63117-631-9

Chapter 1

BUILDING DIGITAL COMMUNITIES: A FRAMEWORK FOR ACTION*

Samantha Becker, Chris Coward, Mike Crandall, Rebecca Sears, Ron Carlee, Kira Hasbargen and Mary Alice Ball

ABSTRACT

Digital inclusion is the ability of individuals and groups to access and use information and communication technologies. Not all members of a community benefit equally, and some communities have been left out altogether. Building Digital Communities: A framework for action is a resource for fostering digital inclusion throughout the United States so that everyone can take advantage of digital technologies. The Framework is structured around a vision for the future, principles that define digital inclusion, goals to make digital inclusion a reality, and sample strategies for achieving the goals.

* This is an edited, reformatted and augmented version of a document released by the Institute of Museum and Library Services in partnership with the University of Washington and International City/County Management Association, dated January 2012.

PREFACE

As access to broadband becomes increasingly important to full participation in society, there is an urgent call for communities to develop technology plans that address the needs of all their residents. Nearly 100 million people in the U.S. do not have access to high speed broadband. Libraries, businesses, hospitals, schools, cultural institutions, community technology centers and local governments have important roles to play and face difficult decisions about how to create and sustain access to broadband technologies and digital content.

In early 2009, Congress directed the Federal Communications Commission to develop a National Broadband Plan to ensure that every American has "access to broadband capability." The National Broadband Plan, released in 2010, noted the significant individual and societal costs of "digital exclusion" and warned that absent action, these costs would grow.

Through its long history of supporting the evolving role of libraries as hubs for community broadband access, the Institute of Museum and Library Services is well positioned to provide leadership in this area. Working together with the University of Washington Information School and the International City/County Management Association we have produced Building Digital Communities: A Framework for Action. The Framework encourages engagement across all sectors of the community so that "all people, businesses, and institutions have access to digital content and technologies that enable them to create and support healthy, prosperous, and cohesive 21st century communities."

The Framework is an important step toward providing tools that can help communities compete in a global economy and improve civic life. It benefited from the input of more than one hundred organizations and individuals with deep knowledge about public access to technology and the diverse information needs of communities. Community and leadership forums were held around the country and a survey was conducted to solicit feedback on the proposed framework.

Libraries and other community-based organizations play especially important roles in meeting the needs of hard to reach residents and ensuring opportunity for all. Libraries are recognized community anchors with the capacity to convene public officials, businesses, schools, hospitals, public safety and cultural institutions so they can work together to fully and effectively realize the value of a digital community.

This report articulates the vision, principles, goals and strategies to help communities realize the benefits that digital technologies offer in economic and workforce development, education, health care, public safety and emergency services, civic engagement, and social connections. The Framework also makes recommendations about how to get started in developing and implementing an action plan for a strong digital community.

We encourage community leaders to use this Framework as a tool to support the development and growth of digitally-connected communities across the nation.

Susan H. Hildreth
Director
Institute of Museum and Library Services

1. INTRODUCTION

What Is Digital Inclusion, and Why Does It Matter?

Digital Inclusion is the ability of individuals and groups to access and use information and communication technologies. Digital inclusion encompasses not only access to the Internet but also the availability of hardware and software; relevant content and services; and training for the digital literacy skills required for effective use of information and communication technologies.[1] The cost of digital exclusion is great. Without access, full participation in nearly every aspect of American society — from economic success and educational achievement, to positive health outcomes and civic engagement — is compromised.

What does digital inclusion mean for people in a community? [2] Simply, it means that:

- All members understand the benefits of advanced information and communication technologies.
- All members have equitable and affordable access to high-speed Internet-connected devices and online content.
- All members can take advantage of the educational, economic, and social opportunities available through these technologies.

The diffusion of the Internet and other information and communication technologies has enabled communities around the United States to reap the rewards of greater connectivity (FCC, 2010). Individuals and organizations from all sectors — including the general public — have been able to expand and enrich their services through these new technologies. Not all members of a community benefit equally, and some communities have been left out altogether. These excluded individuals and communities risk being deprived of basic needs such as education, employment, commerce, and social interaction that increasingly occur through the Internet and other advanced communication technologies.

The obstacles to adoption of communications technologies are varied, as demonstrated in the Summary of Research on Digital Inclusion (Appendix 3). The barriers most commonly cited were individual attitudes to the technology and the cost of access, as well as a perceived lack of relevant content. Several studies pointed to the need for better and wider communication regarding the availability of resources as well as the potential benefits of digital participation.

A Framework for Community-Wide Planning

Communities across the country are recognizing the pressing need to coordinate, plan for, and implement digital inclusion efforts to ensure that every resident is able to benefit in our increasingly digital society. These efforts include extending broadband infrastructure to rural and other underserved areas, digital literacy training to help people adopt digital technologies and educate people about online safety, and targeted programs to ensure everyone can complete an online job application, take a distance learning class, or manage her healthcare.

The Framework is a resource for community-wide planning to help forward-looking communities achieve digital inclusion for all of their residents. It sets forth a vision of what a modern 21st community looks like when digital inclusion is a priority, presents a set of principles that need to be adopted in order to achieve that vision, and provides specific goals that can serve as benchmarks for making progress against each of the principle areas. The Framework also includes strategies that communities have implemented around the country to serve as examples for others, acknowledging that every community is unique and will identify its own direction.

The digital inclusion framework is akin to a transportation planning tool. In transportation planning, communities need to move people and goods in order to prosper and grow. A comprehensive transportation plan includes not just the roads, but also the sidewalks, bike lanes, public transportation, traffic control, and enforcement. A transportation plan uses community-established goals and incentives to prioritize the implementation of improvements, and to respond to the changing needs of the community. Depending on the specific situation of the community (history, geography, financial base, etc.), these priorities are different, but the basic framework behind all of the plans is substantively similar.

So too is it with digital inclusion. Digital inclusion is a requisite for building healthy and prosperous communities across all important sectors—economic and workforce development, education, health care, public safety and emergency services, civic engagement, and social connections. While each community will have different priorities, the fundamental needs are the same:

- for high-speed infrastructure to carry Internet traffic and connect households to businesses and community institutions;
- to help people learn to navigate technology and provide a safe environment for doing so;
- to maintain public options for those who are unable, can't afford, need temporary access, or for whom it doesn't make sense to maintain private access;
- to remove barriers to access for people with disabilities or facing other obstacles to access, and;
- to address the special needs and requirements of key sectors that drive the local economy.

A community must engage all sectors to achieve digital inclusion—this is not just a product of government action, it needs to involve individuals, local and tribal governing bodies, business, the nonprofit community, special interest groups, and other stakeholders. Public libraries, recognizing the role they have played in bridging the digital divide, are uniquely positioned to convene this planning process and begin this essential work, but the initiative can come from anywhere. What is important is that communities raise digital inclusion as a priority and take steps to achieve it. The Framework is a planning resource to help communities in this effort.

Background and Process

In response to the problem of digital exclusion and unequal access, Congress directed the Federal Communications Commission (FCC) to develop a plan to ensure that every American has "access to broadband capability." Issued March 2010, the *National Broadband Plan* recognized the pivotal roles that libraries and community-based organizations play in providing digital training and support, as well as access to high-speed Internet and related community information services, and called on the Institute of Museum and Library Services (IMLS) to

> ...develop guidelines for public access technology based on populations served and organization size.
>
> These guidelines would help libraries and community-based organizations assess their needs for public access workstations, portable devices, and bandwidth. IMLS should work with these organizations to develop guidelines and review them annually to reflect changing technology and practices. (FCC, 2010, Recommendation 9.3.)

To work toward this recommendation, IMLS, working in cooperation with the University of Washington (UW) and its partner the International City/County Management Association (ICMA) (together referred to as the "cooperators") developed a framework ("Framework"), presented in this report, to identify the overarching principles as well as the elements or key characteristics of organizations and communities that foster digital inclusion.

The first step in developing the Framework was to create a draft, or proposed Framework. The preliminary draft of the Framework was the product of an extensive literature review that sought out existing frameworks and digital inclusion efforts implemented in communities across the nation and internationally. Additionally, examples of successful frameworks and indicator efforts were examined to get a sense of how the digital framework should be structured and what elements were necessary to improve its likelihood of adoption.

To refine the Framework and ensure its relevance, over one hundred representatives from libraries, community-based organizations, business, local government, and non-governmental organizations were engaged in the development of the proposed Framework. Their views were sought to help ensure that the Framework will be relevant and useful to professionals in the field of digital inclusion, as well as to increase the credibility of the Framework with communities that will eventually implement it.

The representatives were assembled into two groups. A 16-person Digital Inclusion Working Group (the "Working Group") was selected to assist with the drafting process. The members of this group included a broad range of expertise and represented a variety of organization types, populations, geographies, and technical areas. The Working Group met for two days in January 2011 to review a draft Framework, and also engaged in online discussions and interactive web conferences throughout the Framework development process.

A wider group of about 80 reviewers, known as the Digital Inclusion Network, was formed at the same time, selected by the cooperators with Working Group input. This group represented organizations that are active in implementing digital inclusion policies and services, such as library systems, technology consultants, and educational institutions (see Appendix 2 for a complete list of Working Group and Network participants). The Network provided their comments on draft versions of the Framework through an online social network and also through web conferences.

With the drafting process complete in June 2011, the cooperators prepared to take the proposed Framework to different communities around the country to explore how it could be improved and used by those communities.

Three forums were held in September and October of 2011. The objectives of the forums were to gather meaningful feedback that would support the finalization of the Framework as well as development of implementation and communication plans. IMLS selected three states in which to hold forums – California, Oklahoma, and Maine. The specific cities – Los Angeles, Oklahoma City, and Bangor – were selected with State Library input to provide:

- A mix of rural and urban environments within driving distance of the forum locations
- Income and demographic diversity
- Digital adoption diversity, including populations of non-adopters
- Examples of existing innovative digital inclusion initiatives

Each location hosted two events: a community forum and a leadership forum. The community forum captured information from participants about the nature of digital inclusion in each of the communities, as well as soliciting feedback on the proposed Framework. The leadership forum elicited feedback from community leaders on the Framework as well as on implementation opportunities.

In addition to the forums, IMLS gathered feedback from a wider audience through an online survey. The survey contained questions about the early versions of *Building Digital Communities* materials, as well as attitudinal questions about whether the Framework will support digital inclusion activities.

The resulting final version of the Framework found in this report incorporates insights from the forums and the online survey.

2. Understanding the Framework

Building Digital Communities provides a resource to help communities chart a course toward improving digital inclusiveness — toward expanding the economic and social opportunities provided by digital technology to all its members. The Framework consists of four components:

1) Vision for the future
2) Principles that define digital inclusion
3) Goals to make digital inclusion a reality
4) Strategies for achieving the goals

Figure 1 illustrates the interrelationship of these four components. The foundational vision informs the other three components. The principles yield specific goals. Goals, in turn, shape real-world strategies for implementing the Framework. The sample strategies serve as illustrations of the kinds of initiatives that communities may choose to implement in order to achieve the vision of a digital community.

Four main assumptions informed the vision, principles, goals, and strategies of the Framework:

1) Advanced digital technology enables economic and social well-being.
2) Digital inclusiveness is a worthwhile public policy goal for which public and private resources should be mobilized.
3) Community stakeholders are actively engaged in efforts to build digital inclusiveness.

4) The creation of a digital community requires the involvement of all sectors of the community, any one of which can provide leadership and be a catalyst for action.

Figure 1. Framework structure.

Vision: What Is a Digital Community?

What does a digital community mean for its members? The Framework incorporates vision statements that are forward looking and optimistic about future opportunities enabled by technology. They articulate an overarching digital identity of a community from several perspectives:

- How residents experience a digital community
- How economic growth may be facilitated by technology access
- How government and civic society can be enhanced by connectivity
- How underserved populations might be afforded opportunities through technology

Principles: Statements of Intention

Thirteen principles comprise the backbone of the Framework. Principles are statements of intention that state the rationale for focusing efforts in the areas most important for ensuring digital community now, for planning for the future, and for identifying areas where special effort will be required. The principle statements

- Are fundamental propositions that support specific goals
- Can be considered “mission statements” that further embody the vision statements; and
- Classify aspects of the vision statements into actionable areas.

The thirteen principles are divided into access principles, adoption principles, and application principles. Access principles address the infrastructure a community needs to have in place in order to provide opportunities to benefit from digital life. Adoption principles look to overcome individual barriers that make use of broadband technology less likely, even when access is available. Finally, the application principles look at specific purpose areas where the thoughtful deployment of broadband technologies can enhance the economic success of communities and the lives of their residents.

It is important to note that the ordering of the principles within each grouping is not intended to dictate an overall strategic prioritization; priorities will be determined by individual communities through their self-assessment of community needs. Figure 2 shows the relationship between the three types of principles.

Access principles are:

1) Availability
2) Affordability
3) Design for inclusion
4) Public access

Adoption principles are:

1) Relevance
2) Digital literacy
3) Consumer safety

Application principles are:

1) Economic and workforce development
2) Education
3) Health care
4) Public safety and emergency services
5) Civic engagement
6) Social connections

Figure 2. Access, adoption, and application principles of digital inclusion.

Goals: How to Achieve a Digital Community

Each of the thirteen principles is associated with a series of goals. The goals identified in the Framework articulate the core commitments needed to realize the stated principles and to achieve the vision of a digital community. The goals

- Are tools to help the community focus on achieving the vision
- Point to strategies that enable individuals and organizations to take steps that contribute to achieving the vision statements; and
- Provide a foundation for future benchmarking and evaluation.

Strategies: How the Goals Can Be Achieved

Specific strategies and implementation activities will vary from one community to another depending on local circumstances and priorities. The creation of a digital community requires the involvement of all sectors of the community — and any of them can provide leadership and function as a catalyst for action.

Strategies need to involve four main levels of activity:

A. **Local government, including tribal government,** plays a critical role in all aspects of community planning and development. It has a core responsibility to develop a digital inclusion plan and provide incentives and resources for its implementation as part of its overall strategy to an economically and socially sustainable community.
B. **Libraries, community-based organizations, and other community anchor institutions**[3] are the primary places where people can be assured access to digital technologies, along with training and support. They are safe and trusted institutions, available in most communities.
C. **Businesses** contribute to the building of the broadband infrastructure, as well as developing many of the services enabled through the infrastructure. They both utilize broadband services and are dependent on the readiness of communities to take advantage of the opportunities broadband enables. As part of their business, they may also provide access to the Internet in coffee shops, restaurants, hotels, and other public locations. They play an essential role in providing opportunities for digital inclusion.
D. **Individuals** are at the core of a democratic society. They have the ability to influence community policies, values, and programs as they express themselves in the political process and can advocate for digital inclusion. Digital inclusion, in turn, has the potential to improve individuals' personal, social, and economic well-being and enable them to participate more fully in their communities.

In addition, all stakeholders will need to focus on the broader level of strategic activity:

E. **Influencing policy.** An important dimension of working toward full digital inclusion is advocacy, which requires attention by individual, organizational, and business "champions" at the local, regional, or national level.

The following sections set out the Framework in greater detail. Section 3 briefly discusses the Vision of a digital community. Sections 4-6 outline the access, adoption, and application principles, each with their related goals and sample strategies.

3. The Vision: What Is a Digital Community?

The overarching vision that motivates this Framework is simple but ambitious:

- All people, businesses, and institutions will have access to digital technologies and content that enable them to create and support healthy, prosperous, and cohesive 21st century communities.
- Each community will benefit from harnessing the potential of technology to address its most pressing needs and those of its members.

By choosing to implement the Framework, communities demonstrate commitment to these specific elements of the overall vision:

- Internet access is high speed, affordable, physically accessible, and capable of supporting current demand and future growth.
- Internet access is available to all residents, visitors, businesses, government agencies, libraries, and other community-based organizations (CBOs).
- Public access is available for those who, under normal circumstances or in an emergency, have little or no communication technology at home, or who need to supplement their home connectivity.
- Broader opportunities for economic development are available through full use of information and communication technologies.
- Technology is used to foster social connections, educational and employment opportunities, access to health care, civic participation, and innovation, as well as to drive efficient and effective government services.

- Respectful support is provided for populations that face challenges to participating fully in digital life because of disabilities; physical or cognitive differences; age-related differences in capabilities; language, literacy, and cultural barriers; or other obstacles to using technology.

4. Access Principles

The Access principles include four areas that are the requisite foundation for participation in digital life. Once addressed, they provide the opportunity to benefit from digital inclusion for every individual and community.

Principle 1: Availability

A reliable communications infrastructure is the raw material for innovation, growth, and competitive engagement in today's information-based world. Local and tribal governing bodies, the private sector, and community-based organizations that are bound to one another by geographic proximity and a shared infrastructure need to cooperate to effectively build and maintain appropriate and cost-effective broadband service. Recognizing that greater bandwidth speeds and coverage will be required in the future, these goals are intended to address the immediate needs for high-speed access while preparing for future growth.

Goals

- Access to high speed Internet is available in every household that meets or exceeds the service goals and milestones set by the Federal Communications Commission (FCC).
- Tribal lands and Native communities have affordable access to communication services and broadband technology.
- Sufficient bandwidth is available for multi-user environments, such as educational institutions, public libraries, medical facilities, and other organizations, to support current needs and future network demand.
- Comprehensive standards for broadband readiness in new buildings, renovations, and anchor institutions are adopted by appropriate governing bodies.

- The deployment of infrastructure is guided by right-of-way policies that remove barriers to market entry and system upgrades while limiting disruptions and ensuring other community interests.

Sample Strategies

Individual

- Test actual upload/download speeds against advertised speeds at: http://www.broadband.gov/qualitytest/.

Libraries, CBOs, and Other Community Anchor Institutions

- Ensure that community anchor institutions have access to cooperative high-capacity networks, such as Internet2.
- Develop partnerships with other organizations to share networks and leverage infrastructure investments.

Business Sector

- Support competitive policies to expand markets and encourage innovative options for providing broadband to underserved areas.

Local and Tribal Governing Bodies

- Coordinate efforts and collaborate with other regions to reduce the costs of developing broadband infrastructure.
- Inventory existing network infrastructure, including service quality, costs, and location of community anchor institutions and validate the accuracy of information in the National Broadband Map (http://broadbandmap.gov/).
- Provide strategic incentives to stimulate market competition and private investment in broadband networks in underserved or isolated areas.
- Ensure that local and tribal government agreements with network providers are negotiated to obtain the highest possible download and upload speeds.

Influencing Policy

- Support policies and programs to evaluate the use of wireless spectrum, open additional spectrum, and provide increased bandwidth for use by public institutions.

- Promote standards for measuring upload/download speeds among local Internet Service Providers.

Principle 2: Affordability

Access to the Internet needs to be affordable for communities and their members to benefit from the expanded opportunities of digital life. Public and private entities need to partner together to lower the cost of digital access in general and to provide assistance to those who cannot bear the full cost of home access. Additionally, community members need to be able to understand their options in order to take advantage of competition in the broadband marketplace. These goals are aimed at developing pricing structures and support systems to enable businesses, institutions, and households to afford access to broadband and digital technologies.

Goals

- Internet Service Providers (ISP) provide uniform pricing information to enable consumers to easily compare plans available in the community.
- Programs that subsidize monthly Internet subscription costs are available to low-income households.
- Assistance with hardware, software, and peripheral equipment purchase and maintenance are available to low-income households.

Sample Strategies

Individual

- Donate used technology equipment to nonprofit groups that provide equipment to low-income households and community organizations.

Libraries, CBOs, and Other Community Anchor Institutions

- Raise public awareness about available Internet subsidies and assistance for purchasing computer equipment.

Business Sector

- Donate used computer equipment to nonprofit groups that provide hardware to low-income households and community organizations.

Local and Tribal Governing Bodies

Maintain local comparisons of Internet service provider prices and options.

Influencing Policy

- Promote standards for comparing prices of service bundles among local Internet Service Providers.

Principle 3: Design for Inclusion

People with disabilities, physical or cognitive differences, and differences in age-related capabilities, language, literacy or culture may face challenges using technology and participating in digital life. As a result, they may be denied access to opportunities that could enhance well-being and promote independence. Failure to address the needs of people facing barriers negatively impacts not just those individuals but also the broader community, which will be denied their contributions to its civic, social, and economic health. The following goals reflect a commitment to erasing the boundaries between differences and reducing the obstacles to utilizing technology through principles of effective design.

Goals

- Assistance for ensuring access for people with disabilities and complying, at a minimum, with the letter, intent, and spirit of accessibility laws and regulations is provided to public and private organizations.
- People with disabilities, local and tribal governing bodies, businesses, and community organizations are provided with access to assistive devices and training about the range of technology and design solutions available to accommodate physical or cognitive differences.
- The Seven Principles of Universal Design are embedded into the design of technology-based services for local and tribal governing bodies, businesses, and community organizations. (http://www.ncsu.edu/project/designprojects/udi/center-for-universal-design/the-principles-of-universal-design/)
- Scripting, layout, and navigation of technology based-services are designed, localized, and optimized for multilingual and multicultural use.

Sample Strategies

Individual

- Clearly express personal accessibility needs and preferred accommodations to technology providers, and advocate for these accommodations on behalf of others.
- Participate in volunteer programs that provide one-to-one assistance using technology to people with disabilities.

Libraries, CBOs, and Other Community Anchor Institutions

- Engage in partnerships with community-based organizations that serve people with disabilities to better design services and accommodations.
- Maintain a central resource library of assistive technologies and best practices for accommodating people with disabilities.
- Hold workshops to educate businesses and community organizations about best practices in universal design.

Business Sector

- Align existing online presence with national standards for accessibility and language; move towards universal design for future projects.
- Invest in the development of assistive technology — with the participation of the disabilities community — to reduce cost, improve design, and enable new applications.

Local and Tribal Governing Bodies

- Commit to adopting nationally recognized standards for technology accessibility and language; direct future projects toward universal design principles.
- Assess and manage information technology projects for government and essential services online in a manner that includes people with disabilities and other barriers in all phases of new technology development and deployment.
- Provide education, training, and incentives to businesses and organizations for creating accessible technology and complying with accessibility standards.

- Utilize public information campaigns to make the community more fully aware of the impact of technology design on people with disabilities and language or cultural differences.

Influencing Policy

- Support enforcement of existing accessibility laws as they apply to the digital environment.
- Support national efforts to promote the adoption and use of universal design standards for accessibility.

Principle 4: Public Access

Public access to technology is necessary for community members who have little or no communication technology available in the home, need assistance to effectively use technology, or to supplement connectivity at home or in schools. Others need public access technology to assist during emergencies. Visitors need public access to be able to keep connected to family and work while away from home. Implementing the following goals will ensure that uninterrupted technology and connectivity are available through free public access.

Goals

- The community has sufficient, convenient free access to computers, Internet, wireless networks, and other communication technologies to support the needs of residents, workers, and visitors.
- Public access technology is located in safe facilities, with adequate levels of privacy, security, and accessibility for people with disabilities.
- Information about the availability and location of public access technology is widely disseminated.

Sample Strategies

Individual

- Advocate for sufficient funding for libraries and other community-based organizations to support public access in the community.
- Share stories with policy makers about how public access technology benefits the community.

Libraries, CBOs, and Other Community Anchor Institutions

- Develop partnerships between libraries and community-based organizations to provide public access to technology and technology support in high-need locations.
- Support development of community technology centers for special population groups and low-income housing complexes.
- Provide access to electronic information about community resources and services at strategic locations such as community-based organizations offering social service assistance.

Business Sector

- Develop partnerships and strategic alliances with libraries and other community-based organizations to provide public access equipment, trainers, and other resources.
- Inform users of business-supported wireless networks on best practices for protecting private information on public networks.

Local and Tribal Governing Bodies

- Evaluate the public access needs of the community, develop a public access plan, and provide adequate funding and support for public access in libraries and other community institutions.
- Maintain a directory of public access providers and promote its availability.
- Support the development of wireless hotspots (including public-private partnerships) for residents and non-residents, especially in public spaces such as town centers, parks, and recreation centers.

Influencing Policy

- Support efforts to develop guidelines for determining the levels and types of public access necessary to support a community's needs.
- Support programs that provide resources for local public access, including E-rate and other funding programs.

5. Adoption Principles

The Adoption category includes those principles that are most likely to address individual barriers to adoption. When individuals are uninformed about digital options and opportunities (relevance), don't know how to use digital technologies (digital literacy), or are afraid of using digital technologies (consumer safety), they are less likely to adopt them.

Principle 5: Relevance

Beyond having access to technologies, individuals, businesses, and institutions need to understand the relevance and benefits of using technology to achieve educational, economic, and social goals. Awareness of its potential benefits creates the motivation to master technology skills and learn more about effectively using broadband. These goals aim to help communities raise awareness of the value of technology and generate interest in adopting its use.

Goals

- Community institutions engage in outreach activities to help raise awareness about the benefits of using broadband and other digital technologies among non-adopters.
- Individuals are provided with help to make informed choices about Internet-based services, products, and information.
- Specially designed awareness activities and targeted content are developed to engage older adults in using technology.
- Local and tribal governing bodies, businesses, and community organizations support the creation of multilingual and multicultural content and promote use of digital technologies to diverse users.

Sample Strategies

Individual

- Ask about digital technologies in the government, institutions, and businesses with which one interacts.
- Attend public sessions sponsored by organizations such as the public library to increase one's knowledge.
- If already knowledgeable about digital technologies, volunteer to provide tutoring or training at the public library, youth center, or senior centers.

Libraries, CBOs, and Other Community Anchor Institutions

- Provide training and orientation sessions in multiple languages on digital technologies: what they are, how to use them, and the benefits and cautions around the most common applications.

Business Sector

- Provide information about digital services and products in plain, simple language; create informational material to help people understand their value, and offer one-on-one support to help people understand the technology and products offered.

Local and Tribal Governing Bodies

- Create awareness campaigns about eGovernment and the value of residents accessing local and tribal governing bodies through digital technologies.
- Support the library and other institutions in promoting broad awareness of digital technologies.

Influencing Policy

- Advocate policies that promote plain language explanations of digital technologies to consumers and other users.

Principle 6: Digital Literacy

Digital literacy skills, including the ability to find, evaluate, and use information to achieve goals, are a necessary pathway to digital inclusion. Digital communities meet the needs of their members for learning about technology and maintaining the skills necessary to take advantage of the opportunities enabled by it. The following goals help ensure that communities are able to provide effective digital literacy training to their members.

Goals

- Digital literacy training needs and assets in the community are identified and evaluated, and a strategy for meeting the digital literacy needs of the community is adopted.
- Digital literacy training in multiple languages is provided to individuals, businesses, and institutions through a variety of formats,

including formal classes, real-time virtual help, and one-to-one assistance.

- Digital literacy instruction is embedded in all aspects of curriculum for K-12 and higher education, as well as in lifelong learning activities.
- Information literacy training and assistance is available within the community to help individuals learn to find electronic information and evaluate digital resources.
- Programs aimed at training consumers on the purchase, maintenance, and repair/recovery of technology equipment and services are available through community organizations.

Sample Strategies

Individual

- Volunteer at the library to tutor a new user on basic computer skills.
- Help a neighbor connect to the Internet.

Libraries, CBOs, and Other Community Anchor Institutions

- Coordinate existing efforts to provide digital skills and literacy training between libraries and other community organizations.
- Establish a computer lending program to enable home practice of basic digital literacy skills.
- Organize a "digital literacy corps" of volunteers to improve digital literacy outreach in the community.
- Create a central repository for community training resources.

Business Sector

- Form partnerships with libraries and CBOs to provide trainers for computer skills classes.
- Develop online mentoring programs with local schools.

Local and Tribal Governing Bodies

- Provide online content and services that are designed for all levels of digital skills.

Influencing Policy

- Support strategies to train and provide digital literacy mentors to libraries and CBOs providing community technology services.

- Support the creation of a national outreach and awareness campaign on why broadband matters, focused on key segments of non-adopters.

Principle 7: Consumer Safety

Consumers — both individual and institutional — need accurate, unbiased information on how to safely navigate the digital world. Engaging in digital life requires the knowledge to protect against online threats, and the ability to limit unwanted access to and use of personal information. Online safety is a personal responsibility, but it also requires collective action to educate and assist consumers and enforce standards. The following goals reflect a shared agenda for creating the safest possible online environment for the community.

Goals

- A strategy for training and educating community members about safeguarding personal information, using parental controls, protecting vulnerable populations from online bullying and exploitation, maintaining systems free of viruses, and protecting against other forms of online abuse has been adopted by all providers of Internet services and content.
- Privacy policies are adopted by businesses and government that are visible, easily accessible, and comprehensible to consumers.
- Local law enforcement agencies are equipped with strategies and the authority to pursue cybercriminals while protecting individual civil rights.
- Affordable software and technical assistance are available to support household network security for vulnerable populations.
- Information is widely disseminated through public information campaigns about individual and organizational rights and obligations with regard to intellectual property.

Sample Strategies

Individual

- Maintain awareness of activities children are engaged in online.
- Secure home networks and accounts with strong passwords.
- Educate children about use of technology and online dangers.

Libraries, CBOs, and Other Community Anchor Institutions

- Engage in coordinated educational activities and public awareness campaigns to raise the visibility of online safety issues.
- Develop procedures for securing sensitive client information.
- Develop partnerships and utilize common curriculums to educate consumers on purchasing and maintaining technology.

Business Sector

- Implement policies and processes for protecting customer information and for training staff in following security procedures.
- Support legislation that enables prosecution of cybercrime.

Local and Tribal Governing Bodies

- Provide information about online safety with targeted messages to children and families, the elderly, and other vulnerable populations.
- Fully comply with policies and leading practices to ensure the security and safety of eGovernment activities.
- Ensure that contractors handling sensitive government information have effective safeguards in place to protect data.
- Coordinate efforts with other government agencies to fight security threats to public and government networks.
- Support the investigation and prosecution of cybercrime.

Influencing Policy

- Urge adoption of consumer rights and data protection regulations.
- Support FCC efforts to create a voluntary cybersecurity certification program.

6. Application Principles

The application category includes those principles that represent areas where, once a community and its residents are connected and digitally literate, they can take advantage of the benefits of broadband in real and tangible ways. Community leaders in each of these areas working with the public can best assess local strengths, needs, and opportunities.

Principle 8: Economic and Workforce Development

Communities need to develop the knowledge and skills of future workers and entrepreneurs, as well as help the current workforce update its competencies to meet the needs of employers. Broadband adoption can generate new business opportunities in economically depressed areas and can help ensure sustainable growth. The following goals are aimed at enabling productive public-private partnerships to promote jobs and prosperity, to attract and retain business and workers, and to prepare the current and future workforce to use technology productively.

Goals

- Libraries and community-based organizations engage in partnerships and cross-agency collaborations to enhance the community's workforce training capacity.
- Small businesses and local entrepreneurs are supported by eCommerce and eGovernment tools in their efforts to successfully engage with technology and promote their services online.
- Workforce technology trainers are available in sufficient numbers to support the needs of the community.
- Online career information and specialized assistance with online job-seeking is available through local and tribal governing bodies, libraries, and other community-based organizations.

Sample Strategies

Individual

- Become a mentor or career coach, to help new users of digital technologies understand their potential uses in employment.
- Keep up-to-date with technology skills by attending training provided in the community and through online tutorials.

Libraries, CBOs, and Other Community Anchor Institutions

- Provide assistance, training, tools, and resources for job seekers.
- Develop partnerships with local colleges and technical schools to help new business owners access and use technology.
- Develop public-private partnerships and cross-agency collaborations to better utilize and/or enhance the workforce training capacity of public libraries and community-based organizations.

Business Sector

- Make job application procedures appropriate to the skill level required for the position.
- Provide public kiosks for submitting job applications.
- Develop partnerships with libraries and CBOs to offer workforce training in specialty areas.

Local and Tribal Governing Bodies

- Include digital technologies as part of economic development plans and community economic needs assessments.
- Include digital access in all workforce development and jobs programs.
- Encourage cross-sector partnerships to promote workforce development.

Influencing Policy

- Support the integration of broadband infrastructure into regional planning for economic development.
- Support the creation of innovative national online career tools.
- Encourage federal agency partnerships to foster common goals toward workforce development.

Principle 9: Education

Educational institutions should ensure that students have the digital skills to reach their full potential by connecting them to a diverse range of electronic resources. Technology allows learners to discover educational opportunities, improve academic performance, and prepare to fill the jobs of today and tomorrow. These goals highlight education as a community asset that can benefit from readily available technology.

Goals

- Technology is embedded in curriculum design and instruction, in both formal K-12 and post-secondary institutions to prepare students for 21st century opportunities and challenges.
- Teachers are trained by qualified instructors on the use of technology in instruction.

- Home access to hardware and Internet connections is available for households with K-12 students to support academic success and interactions among caregivers, schools, and instructors.
- Public access technology centers provide assistance for doing homework, conducting online research, completing other academic tasks using digital resources, and supplementing home access.

Sample Strategies

Individual

- Tutor students with their online homework, research, or other academic tasks.
- Volunteer in schools to help others gain digital literacy skills.

Libraries, CBOs, and Other Community Anchor Institutions

- Develop computer loan programs or lease programs to provide learners with computer hardware in the home.
- Provide awareness training for parents on educational technologies being used by students.
- Provide training and support for caregivers to enable them to interact with instructors and schools using digital technologies.
- Coordinate efforts of schools, libraries, and community-based technology centers to maximize delivery of in-school and out-of-school student learning tools.

Business Sector

- Develop relationships with K-12 and post-secondary institutions to provide educators with current information on the skills employers need for the 21st century workforce.
- Create directed learning opportunities for secondary and post-secondary students as a long-term investment in the community's workforce of tomorrow.

Local and Tribal Governing Bodies

- Support the local school system in its digital strategies.
- Leverage cable television and other franchises to support free or reduced cost access to the Internet for students of low-income households.

- Include student access centers in government facilities supporting children and youth, such as recreation centers, libraries, and non-profit youth service providers.

Influencing Policy

- Support efforts to adopt standards for maintaining and securing electronic educational records.
- Support efforts to reduce barriers for obtaining E-rate funding for schools and libraries.

Principle 10: Health Care

Efficiencies and cost-savings in health care delivery, improvement to patient care, and support for independent living and management of health concerns are all enabled by access to broadband and digital technology. Technology supports more effective health care by connecting local practitioners and hospitals with information and specialists around the world; it contributes to better safety for patients by enhancing communication between health care providers and patients; and it creates opportunities for better health education and dissemination of information. The following goals aim to ensure that residents are able to actively manage their own health, and technology is used to provide efficient health service delivery.

Goals

- Secure systems enable local medical professionals and community-based health clinics to share medical records among health care providers.
- Patients have access to user-centered online health information systems, medical records, and private online interaction with health care providers.
- Health care providers and patients are given training on how to use health information systems.
- Transitioning to online health care resources and services is managed to ensure support for new users and non-users of digital technologies, and for those who lack access to technology.

Sample Strategies

Individual

- Volunteer to help patients navigate online information at local hospitals and health clinics.
- Utilize online health resources to improve health, manage chronic illnesses, and communicate with health care providers.

Libraries, CBOs, and Other Community Anchor Institutions

- Provide access to electronic medical records and telehealth services through secure and private public access technology.
- Create consolidated online local health resources accessible through health care agencies and providers, libraries, and other community organizations.
- Provide targeted training for using health technology.
- Provide training for the public on how to locate and evaluate health information.

Local and Tribal Governing Bodies

- Include the use of digital technologies in the development of health and fitness programs in the community.
- Promote partnerships between health care providers and institutions, such as the public library, that provide public access to digital technologies.
- Ensure that public health information is distributed through digital technologies and that adequate access is available to all populations.

Influencing Policy

- Support efforts to establish common standards for sharing research and clinical data.
- Support efforts to ensure that health care delivery locations have access to sufficient bandwidth for high-demand applications.

Principle 11: Public Safety and Emergency Services

Residents and first responders need integrated communications systems for emergency and disaster preparation, response, and recovery. Communities can strengthen their emergency responsiveness through effective deployment of digital technologies: enabling better ways to call for help, coordinate responses, and distribute information during emergencies. Technology can also be mobilized to prevent and investigate criminal activity and to empower residents to improve neighborhood safety. The following goals reflect a need for coordinated, community-wide involvement in planning, preparing, and responding to emergencies and public safety threats in order to create a safe and resilient community.

Goals

- Communities have sufficient secure, resilient, and redundant wireless broadband capacity to support emergency responders throughout planning, preparing, responding, and recovering from emergencies.
- Interoperable and redundant emergency alert networks are in place across mobile, wireless, and wired networks via Common Alerting Protocols.
- Public libraries, schools, and other community institutions provide digital access to residents or evacuees during emergencies.
- Digital tools are used for community-developed online content relating to public safety, emergency preparation, and response.
- Alternative emergency plans account for instances when technology is not available and for how to respond to the needs of community members who do not have technology access.

Sample Strategies

Individual

- Sign up for text and email services offered by local emergency services.
- Create a neighborhood emergency communication network using social media technologies.
- Make a personal plan for emergencies, including contingencies for the loss of digital technologies.

Libraries, CBOs, and Other Community Anchor Institutions

- Engage actively in emergency and disaster planning to identify the range of activities that the institution can support in education, prevention, planning, response, and recovery.

- Develop continuity plans for the institution that provide for resiliency and quick recovery in order to support the community. Specifically create contingency plans for the loss of digital technologies.
- Train staff and test the emergency plans.
- Serve as an information center on the wide range of ways that people can be safe in normal and extreme conditions.

Local and Tribal Governing Bodies

- Address issues of digital access and inclusion in the comprehensive emergency plan. Specifically address the resiliency of digital technologies in the community and make contingency plans for the loss of the technologies and for rapid recovery.
- Actively include the public library and other institutions involved with digital inclusion in planning, training, and testing of emergency plans.
- Make emergency plans, and other public safety information available digitally, with specific emphasis on populations that have special needs.
- Invest in proven current digital technologies such as enhanced 911 systems, computer aided dispatch, and electronic mapping technologies.
- Provide online resources for residents to use for developing personal disaster-preparedness plans.

Influencing Policy

- Support the development of national interoperable public safety wireless broadband networks.
- Support national efforts to ensure that broadband communications are preserved during emergencies.

Principle 12: Civic Engagement

Electronic interaction between community institutions, government agencies, and individuals creates enhanced opportunities for active participation in community affairs. Community members can make use of technology to develop and promote a wide range of civic activities and connections that enrich public life while expanding the network of involved participants. Technology can also help enhance government transparency and accountability. The following goals aim to create increased opportunities for the public to participate in governance, by bridging officials and constituents online and by providing the means for people to voice their opinions and influence decisions.

Goals

- The public connects directly to governments and their agencies, and with each other, in order to learn about and discuss public issues and policies.
- Online access to government information and services meets interoperability standards and is appropriate for users of all skill levels and language needs.
- The community has convenient, secure, and private digital access to government resources and services.
- Technology is utilized to organize community events, to encourage volunteerism and youth participation, and to facilitate problem-solving for community concerns.

Sample Strategies

Individual

- Provide feedback to government agencies about online services.
- Join — or start — community networks to communicate with others about shared concerns.

Libraries, CBOs, and Other Community Anchor Institutions

- Provide a central civic engagement portal for information to residents about local events and resources, as well as online government services.
- Encourage the use of blogs and notification tools so that libraries and CBOs that help people navigate public information can be kept updated with changes to government websites.
- Become "online town halls" for eDemocracy, for participants to shape public agendas and discuss public issues.
- Embed civic engagement and civic knowledge in education and digital literacy training.

Local and Tribal Governing Bodies

- Educate community members about their civic role; provide opportunities for them to interact with government agencies and officials using tools that fit individual or specific community needs.
- Provide residents with the ability to create "My eGovernment" to personalize their interaction with government agencies and officials.
- Encourage the private sector to work with open information, and to develop new applications and ways of using government data.
- Apply care to initial design, updates, and the frequency of redesign to minimize users' confusion and the need to re-familiarize themselves with government websites.

Influencing Policy

- Dedicate resources to develop open standards for government information.
- Promote the adoption of social media technologies that government can use to interact with residents.

Principle 13: Social Connections

Individual members of a community should have access to technologies that promote social engagement and the pursuit of independent learning and creative interests. Technology provides new opportunities for people to express themselves, enjoy activities with family and friends, and get support for personal problems. Lifelong learning and access to new ideas can also spark innovation and creativity. The following goals reflect the ability of technology to foster relationships that support quality of life and well-being in digital communities.

Goals

- Interactive, high-quality multi-cultural content is available through public libraries, museums, archives, and other cultural institutions.
- Programs to encourage development of local content and participation in social networks are available for vulnerable and diverse community members.
- Intergenerational ties are strengthened through technology interaction between youth and older community members.

- The community supports the use of technology for digital preservation and appropriate sharing of local history and contemporary culture in order to build an enhanced sense of community, belonging, and continuity.
- Support and tools are available to produce, archive, and distribute local media programs and other digital content produced by local voices.
- Digital technologies are used as a means of preserving and sharing diverse cultures, if appropriate.

Sample Strategies

Individuals

- Help an older friend or relative connect to online support communities or communicate with distant family members.
- Connect with like-minded people around an interest area or hobby.
- Provide translation services to community organizations serving multilingual populations.

libraries, CBOs, and Other Community Anchor Institutions

- Maintain directories of locally produced content.
- Design a digital story project for collecting local history.
- Support community-based design efforts to mobilize rapid development of high-quality local content by maintaining technology labs and incubators.
- Ensure that online content is available in languages utilized by the community.
- Develop content and applications geared towards a variety of different abilities and cultures.

Local and Tribal Governing Bodies

- Provide funding for cultural institutions to develop local content.

Business Sector

- Use social networking to promote local recreational activities and events.
- Support crowd-sourcing development of new applications.

Influencing Policy

- Advocate for use of social networking to facilitate respectful conversations among people with differing viewpoints.

7. Getting Started on Digital Inclusion

Every community will take its own path to becoming digitally inclusive, depending on its unique set of needs and priorities, the structure of its government and institutions, and its styles of leadership and civic engagement.

Nevertheless, most communities will be able to use a common set of steps to identify both their needs and the actions required to address those needs. The steps outlined below are processes drawn from a range of efforts to build better communities.

1. Convene Stakeholders

Successful strategies to address important community goals are predicated on broad and sustained stakeholder participation, and the list of institutions important to creating a truly inclusive community can be long and diverse:

- The local city, county, and tribal governments, as well as those in neighboring areas, will be essential players. Elected leaders and senior appointed officials have specific responsibilities for strategic and community planning across a wide range of areas.
- Public agencies, especially public libraries, are essential to the process. Whether they are part of local or district government or independent institutions, public libraries have become the central institutional player in providing public access to digital technologies. Other important public agencies include the public schools (K-12), institutions of higher learning (especially community colleges and adult education programs), economic development agencies, agencies with community centers (such as recreation departments), and public housing departments. In many local governments a central office of technology plays a critical role in digital management issues across the various public agencies and should be actively engaged in the discussion.

- Non-profit community-based organizations can be especially effective in reaching hard-to-serve populations, and vary widely from one community to another. Community-based organizations are often more nimble and experimental than government entities. Likely partners for community-based organizations are non-profits that serve targeted populations within a community — including people who are low-income or homeless, populations with disabilities, or people from specific ethnic cultures. Organizations that provide supportive housing, job training, and childcare can reach residents who need access to digital technologies and that may have otherwise been overlooked. Community foundations can serve as neutral conveners and as sources of initial funding.
- The business community includes many potential partners. Telecommunication companies have an essential role. Major employers in need of trained employees can provide valued support, especially if there is a single, dominant employer within the community. Chambers of Commerce and other business groups often recognize the economic imperatives for digital inclusion. Businesses play an obvious role in the aspects of digital inclusion impacting economic development. Many businesses make a point of incorporating all segments of the community in their civic and social philanthropic activities. Businesses can also provide Internet hotspots for public access.
- Residents, be they individuals or representatives of neighborhood or housing associations, can help ground digital inclusion efforts in the needs that are most relevant to the community. Civic engagement begins with becoming informed on digital inclusion issues, and is followed by engaging with local organizations in order to raise peoples' awareness about the importance of adopting broadband and actively participating in a digital society.

  Residents' participation on committees and task forces help institutions develop a broader perspective of the values of digital inclusion and its effect on the community; institutional and organizational leaders can expand resident participation by conducting surveys, forums, and workshops in the communities where they live and for the people they serve.

Such a diverse set of stakeholders will not inevitably come together. Participatory action will require leadership — a catalyst to initiate the participatory process, engage the broad array of stakeholders, and guide the process to success. Given its experience in digital technologies and its reputation for being non-partisan and welcoming to all, the public library is uniquely positioned to assume this leadership role.

2. Develop a Shared Community Understanding of Digital Inclusion

In an area as complex as digital inclusion, community stakeholders need to begin with a shared understanding of the needs and the goals.

- What does the term digital literacy mean for the community?
- What digital technologies are currently available, and to whom?
- Where are the gaps? Who is left out and at risk of being left behind?
- What are the most important community goals of digital inclusion: economic development, education, job training, health care, emergency management, social connection?

The answers to these questions will necessarily vary by community. A key component of developing community answers should be a needs assessment that is based on systematic data collection and analysis. A shared vision should emerge from the joint learning process – a vision that describes where the community wants to be in a technology-driven world and what it needs to do to get there.

3. Create a Community Action Plan

Grounded in the shared vision, communities need to develop an actionable plan that outlines specific goals, measures of success, timelines, and assignments of responsibility. Some communities will make large comprehensive plans; others will create more narrowly targeted plans that utilize their limited resources as effectively as possible. Either approach can work.

4. Implement the Plan

Drawing on government appropriations, business contributions, philanthropy, or combination of sources, communities will need to develop the mechanisms necessary to generate the resources required to execute the plan. It is important to ensure early success that can be celebrated and promoted, creating awareness and building momentum. A well-structured monitoring effort will allow community stakeholders to provide mutual support, encouragement, and accountability.

5. Evaluate and Revise the Plan

Change is the only constant. Digital devices and transmission capabilities, workforce requirements, economic drivers, and population demographics are all constantly changing. Change may be incremental and easily incorporated. Or, change may be shockingly dramatic, as when a major employer leaves – or one arrives. As circumstances change, implementation plans must change as well.

In today's information-based economy, digital inclusion is not a simple, one-time checkbox. Digital inclusion will require sustained effort, ongoing evaluation, and the willingness to revise the community's plans and strategies.

APPENDIX 1. PARTNER INFORMATION

Institute of Museum and Library Services

The Institute of Museum and Library Services is the primary source of federal support for the nation's 123,000 libraries and 17,500 museums. The Institute's mission is to inspire libraries and museums to advance innovation, lifelong learning and cultural and civic engagement. The Institute works at the national level and in coordination with state and local organizations to provide leadership through research, policy development and grant-making. To learn more about the Institute, please visit www.imls.gov.

University of Washington

The Technology & Social Change Group (TASCHA) at the University of Washington Information School explores the design, use, and effects of information and communication technologies in communities facing social and economic challenges. With experience in 50 countries, TASCHA brings together a multidisciplinary network of social scientists, engineers, and development practitioners to conduct research, advance knowledge, create public resources, and improve policy and program design. Our purpose? To spark innovation and opportunities for those who need it most. Learn more at tascha.uw.edu.

International City/County Management Association

ICMA, the International City/County Management Association, advances professional local government worldwide. Our mission is to create excellence in local governance by developing and advancing professional management to create sustainable communities that improve lives worldwide. ICMA provides member support; publications; data and information; peer and results-oriented assistance; and training and professional development to nearly 9,000 city, town, and county experts and other individuals and organizations throughout the world. The management decisions made by ICMA's members affect millions of individuals living in thousands of communities, from small villages and towns to large metropolitan areas.

Appendix 2. Partner Team Members

Digital Inclusion Working Group

Steve Albertson	Community Voice Mail
Mary Carr	Spokane Community College
Mark Cooper	Consumer Federation of America
Cathy DeRosa	OCLC
Jon Gant	University of Illinois
Chris Gates	Philanthropy for Active Civic Engagement (PACE)
Martín Gómez	Los Angeles Public Library

Link Hoewing	Verizon
John Horrigan	TechNet
David Keyes	City of Seattle
Mike Lee	AARP
Traci Morris	Homahota Consulting
Mare Parker O'Toole	Medfield Public Library
Frances Roehm	Skokie Public Library
Jane Smith Patterson	e-NC Authority
Sarah Washburn	TechSoup

Digital Inclusion Network

4-H National Headquarters	Jim Kahler
Access Humboldt	Sean McLaughlin
Access to Justice	Don Horowitz
Alaska Children's Trust / Friends of the Alaska Children's Trust	Panu Lucier
Alcatel-Lucent	Greg Kovich
American Association of Museums	Zeinab Ulucan
American Foundation for the Blind	Paul Schroeder
American Library Association	Alan Inouye
Arizona Telemedicine Program	Mike Holcomb
Association of African American Museums	Bill Billingsley
Balboa Park Online Collaborative	Perian Sully
Bill & Melinda Gates Foundation	Karen Archer Perry
Blacksburg Electronic Village	Brenda van Gelder
Blandin Foundation / Treacy Information Services	Ann Treacy
Boys & Girls Clubs of Tennessee	John K. Berry
California Academy of Sciences	Elizabeth Babcock
California State Library	Stacey Aldrich
Center for Media Justice	Amalia Deloney
Cleveland Public Library	Felton Thomas
Comcast	Juan Otero
Communication Service for the Deaf	Benjamin J. Soukup
Computers for YouthElizabeth Stock	
Council of State Governments	John Mountjoy
Dell	Kerry Murray

Department of Labor, Center for Faith-based & Neighborhood Partnerships	Rev. Phil Tom
E-Democracy.org	Steve Clift
EdLab	Karen Peterson
Georgetown County Library	Dwight McInvaill
Georgia Division of Archives and History	David Carmicheal
Global Center for Cultural Entrepreneurship	Alice Loy
Google	Jenn Taylor
Grand Rapids Community Media Center	Laurie Cirivello
Greater Auburn-Gresham Development Corporation	Ernest Sanders
Hardy Telecommunications	Derek Barr
Indianapolis Museum of Art	Ron Stein
Instructional Technology Council	Christine Mullins
Intel	Rick Herrmann
Iowa Statewide Interoperable Communications System Board	Jim Bogner
J. Paul Getty Museum	Nik Honeysett
Knight Foundation	Judith Kleinberg
Lyrasis	Kate Nevins
Maine State Library	Linda Lord
Miami Dade Public Library System	Raymond Santiago
Microsoft	Andrea Taylor
NAACP	Hilary Shelton
National Association of Counties	Jackie Byers
National Indian Telecommunications Institute	Karen Buller
National Information Standards Organization	Todd Carpenter
National Internet2 K20 Initiative	James Werle
National Library Service for the Blind and Physically Handicapped	Jennifer Sutton
Net Literacy	Don Kent
New York State Library	Mary Linda Todd
Nez Perce Tribe	Christina St. Germaine
Nonprofit Technology Network (NTEN)	Holly Ross
Oklahoma State Library	Susan McVey
Partnership for a Connected Illinois	Debbie Strauss

Pew Internet & American Life Project	Lee Rainie
Public Library Association	Barb Macikas
Rainbow PUSH Coalition	Kimberly Marcus
Rhode Island Economic Development Corporation	Stu Freiman
Rural School and Community Trust	Rachel Tompkins
Rural Telecommunications Congress	Greg Laudeman
Saint Paul Neighborhood Network	Mike Wassenaar
Saint Regis Mohawk Tribal Council	Jamie Bay
ShinyDoor	Angela Siefer
Syracuse University	David Lankes
Telecommunity Resource Center	Gene Crick
Texas State Library	Peggy Rudd
Texas State Library and Archives Commission	Chris Jowaisas
Transmission Project	Belinda Rawlins
University of Illinois	Colin Rhinesmith
University of Maryland	John Bertot
University of Texas at Austin	Sharon Strover
University of Washington	Miranda Belarde-Lewis
University of Wyoming	W. Reed Scull
Urban Libraries Council	Susan Benton
VisionTech360	Bill Gillis
Weeksville Heritage Center	Pamela Green
Youth Policy Institute	Sarah Serota

APPENDIX 3. SUMMARY OF RESEARCH ON DIGITAL INCLUSION

Internet connectivity offers a range of important benefits for individuals and communities. It increases equal access to opportunities such as jobs and work force training, enables people to find information about their health, spurs civil engagement, and supports other productive activities. Recognizing the Internet's growing ubiquity, there have been many efforts to reduce the barriers to access to information and communication technologies (ICTs). Initially the focus was on removing obvious barriers to access — a technology-centric approach concerned mainly with lowering the cost of entry into the digital marketplace. Lately there is growing recognition that the more significant barriers include such factors as attitudes, interests, and abilities.

In view of this shift in understanding, current initiatives — instead of aiming to bridge a "digital divide" — are increasingly directed toward fostering "digital inclusion." The technology-centric term digital divide separates people into haves and have-nots; digital inclusion is an all-encompassing term, focused on equity within three areas: access, digital and information literacy, and relevant content. Digital inclusion recognizes that facilitating participation in digital life is an objective that will require ongoing, thoughtful action, including making available the necessary technology. This people-centric perspective will accordingly call for an ongoing assessment of community conditions and needs. *Building Digital Communities: A framework for action* ("Framework") aims to be a tool for such assessment.

This section provides an overview of the articles, reports, and other resources that informed the creation of the Framework. The review covers materials in four subject areas:

1) Self-assessment models
2) Indicator systems
3) Local and international digital inclusion policy initiatives
4) Reasons for non-adoption (and the unmet needs of those without digital access)

Assessment Models

Three main assessment models helped inform the structure of the Framework: the LEED Certification Rating System (U.S. Green Building Council), Tree City USA (Arbor Day Foundation), and the National Information Standards Organization (NISO) Framework of Guidance for Building Good Digital Collections. These models share a number of design strengths. They are aspirational, flexible, and easy to understand, enabling grassroots action towards specific goals and also fostering collaboration among organizations. That is, these models emphasize a "bottom-up" approach.

More specifically, each model provides a strong overall purpose, an easily comprehensible structure and categories of focus, and specific guidelines for meeting the overall goals of the model. They differ in levels of prescriptiveness, elaborateness, and technical specifications. For example, the

LEED system has strict requirements for each level of certification, and applicants are able to "self-select" which level they feel qualified for and choose among a menu of options to achieve it. In contrast, the Tree City USA certification system has just four criteria, which are straightforward and relatively easy to achieve. *Building Digital Communities: A framework for action* aims to provide a clear structure and targeted goals that allows for flexibility in implementing strategies.

Indicator Systems

Indicator systems provide useful guidance for achieving the right level of specificity for the Framework; they also help inform how to measure progress across many different types of communities. Shared indicators are used widely in social, economic, environmental, health, and educational policy domains. They are usually developed by a government or nonprofit sponsoring organization, typically beginning with a process of soliciting community input in determining the indicators to track. An example is the Boston Indicators Project — a set of tools for organizations to track trends and measure progress towards such community goals as improved health, greater literacy, and reduced poverty.

Indicator systems have many advantages; they allow shared data collection and can produce useful quantitative data on progress toward community goals. However, the level of specificity in these systems was considered to be beyond the scope of this Framework. Communities will need to begin by developing their own working definition of digital inclusiveness and by setting their own goals for achieving that objective — and developing indicators is a later step.

Nevertheless, the participatory process used in developing indicator systems was useful in guiding the process of developing the Framework. For example, the Burlington Legacy Project engaged citizens of Burlington, Vermont in an inclusive "listening and learning" process to set sustainability goals for the city's future. Other indicator systems relied on experts and representatives from different sectors to help shape the structure and content of the indicators before validating them with a community. Work on the proposed Framework combined these two approaches. A diverse group of experts from many different sectors assembled to discuss all of the elements, engaging in a serial review process as the proposed Framework was developed. In addition, the proposed Framework was introduced to select communities as part of a series of forums to hear from the general public and from individuals and organizations who would be responsible for implementing it — in a "teaching, listening, and learning" process.

Local and International Policy Initiatives

The initial elements of the Framework — the vision, principles and goals — emerged from an examination of local and international policy initiatives around digital inclusion. Although each digital inclusion policy initiative is shaped by the specific concerns of the particular community, all share many common goals and strategies. Most commonly, they highlight the themes of access, relevant content, and information literacy; other shared areas of concern are infrastructure, civil engagement, and affordability.

- Access goals relate to the software and hardware that enables connectivity; they may also relate to affordability and infrastructure.
- Providing local information for and by local residents is part of relevant content, and is considered important for creating interest in digital life; it is also an aspect of eGovernment as well as civic engagement in policy initiatives.
- Affordability refers to costs, usually to individual users of technology.

The definitions of infrastructure and information literacy vary across policy initiatives. Infrastructure initiatives in the U.S. have — until recently — been primarily at the state or local level; they set network capacity goals based on existing local conditions. Non-U.S. initiatives often set more ambitious infrastructure targets, as in bringing broadband or wireless connectivity to rural populations. Information literacy comprises a wide range of definitions rather than a standardized skill-set. For example, U.K. information literacy efforts focus on job-related applications, whereas U.S. efforts emphasize the skills needed to conduct job searches on the Internet. In San Francisco, basic information literacy includes both these objectives. (Indeed, San Francisco's ambitious initiative, inspired by its nearby technological hub, was designed to enable its citizens to become creators and innovators of Internet content as the long-term goal of information literacy.) Most definitions of information literacy tend to overlook the growing interest in the "fun" aspects of Internet connectivity — although surveys indicate that people want to participate in online social networks, including playing games. More recent discussions embrace these and other expansive understandings of information literacy.

In developing the Framework, the common focus areas and goals were gleaned from these different initiatives. Important goals also came from the National Broadband Plan, addressing inequalities in access, planning for future expansion, and building social and economic well-being at the local level. The goals were further shaped to reflect a forward-looking, optimistic vision of how a digital community may be experienced by its residents.

Reasons for Non-Adoption

The explosive growth in the use of information and communication technologies in everyday life, while offering many positive benefits, has also left many Americans behind. Research shows that technology adoption and digital literacy levels differ according to attitudes about the Internet, income, education level, disability, and age. In gathering information for the Framework, an important element was to incorporate an understanding of non-adopters, embedding goals that address their concerns.

There are a number of barriers to adoption. Cost is a major barrier: low-income homes are less likely to have at-home access to the Internet, due in part to high Internet subscription rates. Hardware and software costs can also result in non-adoption, or cause users to become "digital drop-outs." The elderly and people with disabilities are also less likely to have at-home Internet access or to use public access computers. To address these barriers, research documents three strategies: literacy classes, tiered (or reduced) Internet subscription rates, and outreach efforts.

Classes for digital literacy (covering computer ownership and maintenance) help new users gain confidence as well as skills; more advanced classes that introduce specific applications provide opportunities for embedding new information literacy skills. Reduced, subsidized, or tiered subscription rates make access affordable at home (and at community institutions).

Creating or improving outreach is important especially to publicize such lower-cost Internet options, digital literacy programs, and related digital inclusion efforts. In many initiatives, implementation has included outreach in the earlier stages of the process, inviting residents and business to articulate from the start what specifically matters to them.

According to many studies, however, the primary reason for non-adoption is not cost but lack of interest. Research distinguishes several groups, based on different attitudes: those who do not recognize Internet access as a need; those who see access as a need, but who face financial barriers; and those who are inhibited by non-financial barriers. In order to counteract lack of confidence, fear, and mistrust, the benefits of connectivity need to be properly communicated, by advocating for information and digital literacy and by helping people understand the significance of being able to do things online.

Lack of relevant content is another reason for non-adoption. Research suggests that if non-adopters fail to "see" a place for themselves on the Internet, with content that is relevant to them, their interest wanes. For example, some non-English speakers cannot find materials to read online. Even when language is not a barrier, local content may be lacking. One article reported that people searching for ads for local businesses were unable to locate that content. There is a basic need to connect offline experiences to online content.

In rural areas that currently rely on dial-up access, adoption continues to be a problem. Because telecommunications providers lack a financial incentive to upgrade their services, and public funding is inadequate to spur infrastructure investments, many eager broadband adopters are left wanting and are effectively excluded from digital life because dial-up speeds are no longer adequate to be considered "true access."

Finally, the hardest to reach populations are the elderly, the poor, and the less educated, followed by racial or ethnic minorities and people with disabilities. Targeted outreach to these populations is essential.

These differences in reasons for non-adoption, along with the recognition that building digital literacy skills happens over a long period of time, lie behind many of the goals in the Framework. Examples include the focus on providing digital and information literacy instruction, and ensuring that transitions to online applications (as in eGovernment and Health) take into account the needs of new users of digital technologies, people with disabilities, and others with barriers to access. The Framework also emphasizes the need to build awareness of the benefits of connectivity (the demand side), in addition to supplying infrastructure and providing more access points (the supply side). Greater demand will ensure broader use of improved resources and sustained interest in digital inclusion initiatives.

Conclusion

The review of self-assessment models, indicator systems, public policy initiatives, and digital inclusion research covered over 240 resources (see Bibliography). This extensive review helped define the structure and scope of the Framework, while informing its specific principles, goals and strategies.

APPENDIX 4. OVERVIEW OF COMMUNITY AND LEADERSHIP FORUMS

IMLS, UW, and ICMA convened three community and three leadership forums for digital inclusion in September and October 2011. The forums generated feedback from community residents and leaders on the general topic of digital inclusion, the content and structure of the Proposed Framework, and the opportunities and challenges for implementation. The first took place on September 12 in Los Angeles, with additional events held on October 11 in Oklahoma City and October 18 in Bangor, Maine. The three states in which to hold forums – California, Oklahoma, and Maine – were selected to provide a level of national regional diversity, in addition to existing broadband initiatives and strong leadership by the state library. The specific cities – Los Angeles, Oklahoma City, and Bangor - were selected with State Library input to provide:

- A mix of rural and urban environments within driving distance of the forum locations
- Income and demographic diversity; mix of digital saturation and pockets of non-adoption
- Existing innovative digital inclusion initiatives or close connections with other organizations working in the digital inclusion arena

IMLS, UW, and ICMA partnered with the state libraries of California, Oklahoma, and Maine to organize the events. These libraries – with financial support from the Bill & Melinda Gates Foundation - worked with local libraries in their home states to host each community and leadership forum, including the selection of attendees.

Community Forums

The community forums began with a presentation and plenary session to provide basic education on digital inclusion, the structure and content of the Proposed Framework, and to frame small group discussions. The plenary session included a panel discussion where the local library director discussed local broadband efforts and the relevance of digital inclusion in that location.

Later, attendees were separated into four discussion groups. Small group moderators from the UW lead the groups in a process of deliberative dialogue whereby participants began conversation with their personal experiences or digital inclusion (or exclusion) and proceeded to exchange ideas about digital inclusion in their communities and how the Framework might be implemented to improve digital opportunities.

At the end of the evening, discussion groups were asked to "report out" the most important themes of their discussion to the entire community forum. In many cases, the themes were not principles or goals at all, and rather reflections on the flow of the conversation or next steps. For example, one group in Oklahoma reported that they spent very little of their time together talking about devices or technology, and a great deal of it speaking about people and partnerships.

Description of Attendees Across Three Locations

A total of 115 residents of areas in and around Los Angeles, Oklahoma City, and Maine attended the community forums. Male and female participants, who were selected based on their affiliation with targeted community groups, background, and interest in digital inclusion, ranged in age and professional experience. They did not have to have prior knowledge of the Framework for Digitally Inclusive Communities to be invited to attend.

- In Los Angeles, the community forum was attended by 37 people who came from an art museum, cultural heritage groups, the Los Angeles Public Library (e.g. docents, librarians, and directors), the Los Angeles Unified School District (e.g. administrators, teachers, and parent support services), various Friends groups, social services organizations, and technology education centers.
- In Oklahoma City, the community forum was attended by 31 people who represented Asian ethnic groups, city government, Native American tribes, K-12 education, community colleges, cultural

museums, health care, people with disabilities, public libraries, public radio, and social services organizations.

- In Bangor, the community forum was attended by more than 40 people such as city government officials (e.g. city manager, mayors, and the chief of police), disability aides, K-12 educators, health professionals (e.g. nurses and Medicare liaisons), information technology workers, librarians and library staff, local business owners, students, and volunteers.

Specifically, Cooperators sought feedback on underserved populations including people from rural areas.

Description of Data Collection and Analytic Methods Used

UW used multiple methods to collect data about participants' experience with and feelings toward digital inclusion and the proposed Framework for Digitally Inclusive Communities throughout the forum, including the discussion groups, polling, and surveys.

Discussion Groups

Discussion groups were recorded and notes were taken by moderators and volunteer scribes. The data were coded using two approaches and combined for cross-tabulation:

1) Where participant dialogue was directly related to specific principles and goals, the dialogue was labeled using the Framework as the source for codes. This approach was used to examine the frequency of comments made about specific principles.
2) General themes were identified in participant dialogue. The source of the coding was the dialogue itself, rather than the Framework. This approach was used to assess whether or not the Framework captures the most significant concerns of digital inclusiveness as expressed by community members.

Polling

During the small group discussion, participants were given three colored dots and asked to use them to identify the most important principles in the Framework. Participants were able to use as many of their dots as they wished on one principle, providing them with the opportunity to "bullet vote" for principles they felt were highly important relative to others.

Surveys

Attendees were asked to complete a survey upon arrival and another survey at the end of the session, in order to analyze attitudinal changes among attendees between arrival and departure. Both surveys contained the following questions:

1) How knowledgeable do you feel about digital inclusion?
2) How important of an issue is digital inclusion for you personally?
3) How important do you think it is for your community to work towards digital inclusion?
4) If digital inclusion was made a priority in your community, how much do you think economic opportunities would improve?
5) What are the 3 most important principles of digital inclusion for your community?

Leadership Forums

Leadership forums were conducted the day following the community forums with the following participation: Los Angeles (5 participants), Oklahoma City (11 participants), and Bangor, Maine (30 participants). The leadership forums allowed direct engagement with the community leaders best positioned to encourage and organize the implementation of the Proposed Framework.

The leadership forums were intended to convene leaders in a community who had the position and authority to advance digital inclusion efforts. Leaders were asked if they found the framework helpful and whether or not they would use it in their community. Additionally, they provided feedback about implementation approaches.

Leadership forum participants included library leaders, local government representatives (appointed and elected), and non-profit and/or business leadership that influence local policy priorities. Participants in the leadership forums were selected to represent the demographics and specific concerns of communities. Attendees were selected by ICMA and the state librarians of California, Maine, and Oklahoma, with final approval by IMLS. Invitations were sent from the state librarian, and in the case of local government invitees, ICMA sent follow-up invitations.

Format of Leadership Forums

The leadership forums focused on the observations of the group as they related to the principles and goals of the framework, implementation options, potential obstacles to implementation, and the tools needed to implement the framework in their communities. To gain a better understanding of the digital inclusion landscape in their communities and to hear the needs of local residents, leadership forum participants were strongly encouraged to attend and observe the community forums.

Except for the Bangor, ME leadership forum where parts of the forum were conducted in smaller groups, the leadership forums were conducted as open group discussions, wherein attendees were encouraged to respond to directed questions, ask questions of each other, and ask questions to the Cooperators Team.

Directed questions were kept high level. The questions posed to attendees in the leadership forums included:

1) What are your thoughts on digital inclusion?
2) Would a national framework be useful?
3) What do you think about the framework's structure, the principles and goals?
4) What's missing from the framework?
5) What implementation scenario would be the most helpful to you?
6) What tools would you need to adopt and implement the framework?
7) How could we make the framework more relevant to the work you're doing?
8) What advice do you have for us?

APPENDIX 5. OVERALL ANALYSIS OF THE COMMUNITY FORUMS

Summary of Overall Analysis of the Community Forums

- While only 11% of community forum attendees felt very knowledgeable of digital inclusion before attending the forum, 29% felt very knowledgeable after participating in the forum.

- 17% of the participants felt that digital inclusion was a very important personal issue before the forum, while after participating in the forum 60% felt that digital inclusion was an important personal issue.
- Even before attending the forum, 64% of participants felt that their community should be addressing issues of digital inclusion; after participating in the forum, 72% felt that their community should be addressing digital inclusion issues.
- The survey respondents selected the following principles most frequently: availability & affordability, education, economic & workforce development and public access.
- Over the course of three community forums and twelve discussion groups, there were a total of 268 documented comments about availability & affordability. The next three most discussed principles were adoption & digital literacy, quality of life, and education.

Surveys & Principle-Specific Findings

The surveys and polls conducted during the community forums showed a strong trend to changing attitudes as a result of participating in the forums and discussions with others about the topic of digital inclusion. Overall, participants felt they were more knowledgeable about digital inclusion, and also felt more strongly that digital inclusion is an important issue for themselves, personally, as well as for their communities.

Figure 1 shows the change in knowledge levels about digital inclusion. The intake survey found that 11% of attendees felt *very knowledgeable* of digital inclusion before attending the community forums, while the exit survey found that after participating in the forum, 29% then felt *very knowledgeable* of digital inclusion.

From the intake survey to the exit survey, the number of attendees who felt that digital inclusion was a very important personal issue grew by 17% to 60%. In both surveys, the majority of participants felt that it was *very important* for their community to work towards digital inclusion, however, even this attitude shifted towards a perception of even greater importance among some participation.

To help understand whether the Proposed Framework and subsequent implementation activities would be perceived as valuable for communities, attendees were asked if they believed economic opportunities would improve if digital inclusion was a priority in their community. Before attending the forum, 43% of participants responded that they believed economic opportunity would improve *a lot*; by the end of the forum, that perception had increased to 50% of attendees feeling economic opportunities would improve *a lot* if digital inclusion was a priority in their community.

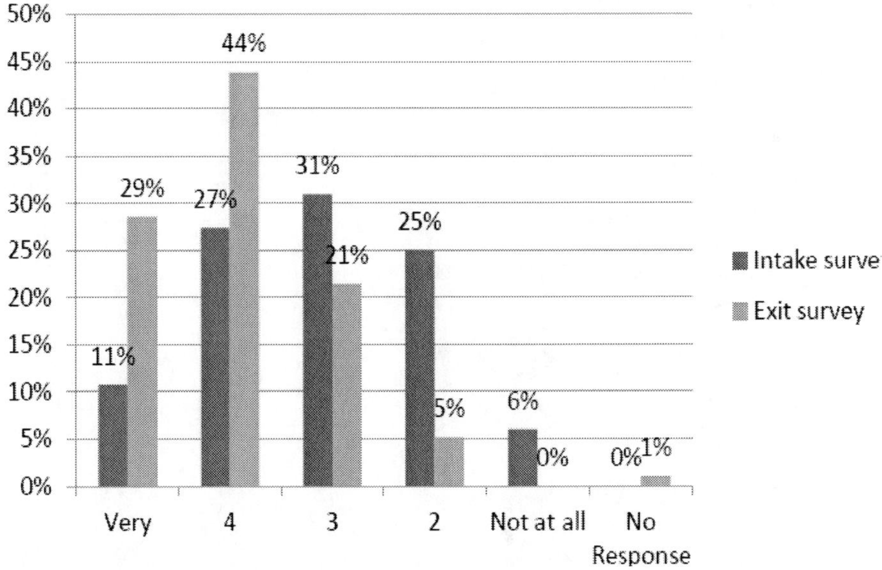

Figure 1. Knowledge of digital inclusion.

In the forced choice exercise presented in the intake and exit surveys, attendees were asked to choose the top 3 most important principles in the Proposed Framework principles. At polling during the small group discussions, attendees were allowed to apply three votes to the principles, giving them the ability to vote more than once for principles they thought were especially important. Before engaging in discussion, public access, education, and economic & workforce development were considered the most important principle areas. After discussion with their small group, however, the availability & affordability rose to the most important place, while education, and economic & workforce remained in the top three. Figure 2 below demonstrates the side-by-side ranking of Framework principles from the intake survey, exit survey, and the poll.

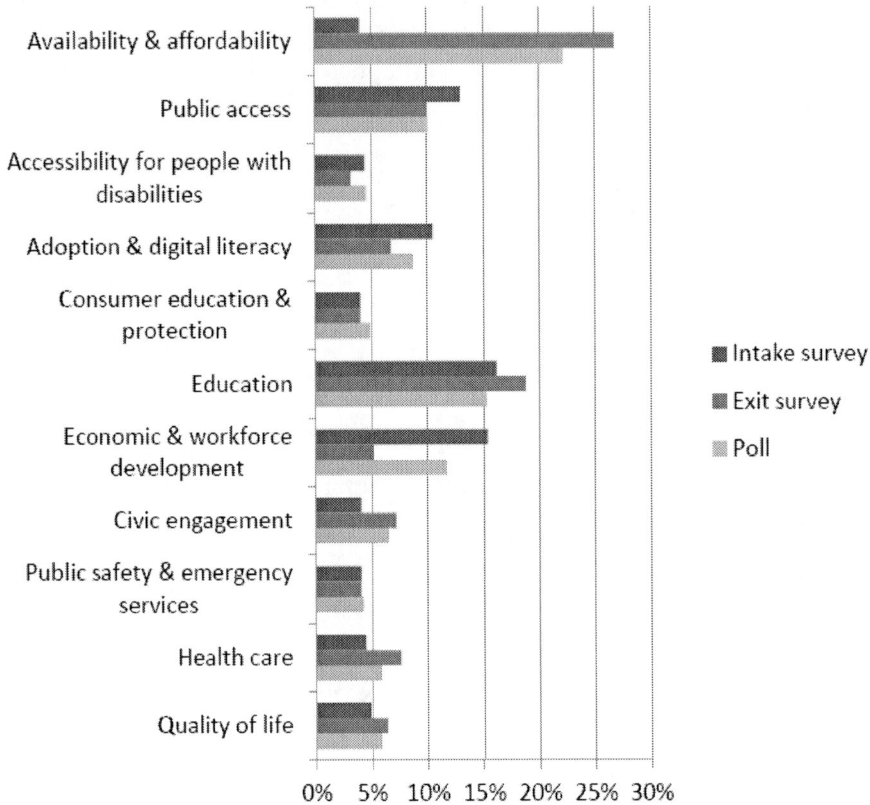

Figure 2. Ranking of Framework principles from surveys.

Within the groups, certain principle areas garnered more discussion. Consistent with the ranking exercise, availability & affordability was the most discussed aspect of digital inclusion. The next three most discussed digital inclusion principles were adoption & digital literacy, quality of life, and education.

Conversations about the principle areas somewhat varied by city. For example, though it received 31 unique comments in Bangor, the four discussion groups in Los Angeles spent little to no time discussing health care. Oklahoma City was more concerned with adoption and digital literacy than the other locations.

Discussion Themes

For each community forum site, general themes emerged both related to the Proposed Framework and arising spontaneously from the discussion groups. These themes were ranked across the forums and classified as high-range, midrange, or low-range. High-range themes appeared in all four discussion groups within a single forum, mid-range themes in three out of four of those discussion groups, and low-range themes in two out of four of the discussion groups.

Table 1 shows the cumulative themes, how the themes were ranked at each site, and common ranking among two or more sites. No common themes among the three cities shared a common ranking. However, many themes overlapped between two locations.

Table 1. Discussion themes by number of sites and ranking (high, medium, or low)

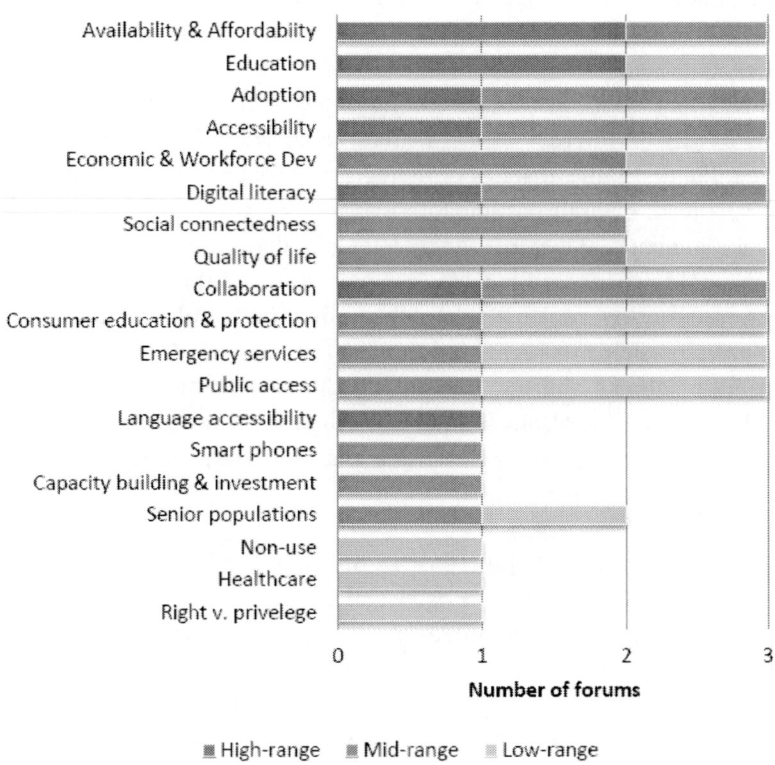

High-Range Themes – Present in at Least 10 of 12 Discussion Groups

Availability and Affordability

Availability and affordability was discussed in 11 of 12 discussion groups, the only theme with this distinction. It was a high-range theme in Los Angeles and Oklahoma City and a mid-range theme in Bangor. In Los Angeles, participants addressed the issue of reliable speed and affordability. Participants expressed frustration with inconsistent bandwidth speed and a disparate range of ISP prices. In both Oklahoma City and Bangor, attendees expressed the need for broadband technology infrastructure in rural areas. Both cities recognized the difficulty in motivating ISP providers to expand to rural towns, too. Attendees from the Oklahoma City forum shared that there are more options available in rural areas now, but still the prices are too high for residents.

Education

Education was discussed in 10 of 12 discussion groups. Education was a high-range theme for both Oklahoma City and Bangor. In Oklahoma City, attendees spoke about the lack of digital resources in their public schools and recognized that some schools have no Internet access at all, forcing their students to rely on Internet access at libraries instead of their own schools. On the other hand, Maine offers computer and broadband access in all public schools. Instead, Bangor participants worried about the lack of home Internet access for students. They recognized that students are disadvantaged when they are unable to use online educational tools at home after school hours. Education was a low-range theme in Los Angeles. Similar to Oklahoma City, Los Angeles participants addressed the disparity in digital resources among the city's public schools and the lack of broadband infrastructure in many older school buildings. Both Bangor and Los Angeles attendees hoped to broaden the principle to include lifelong learning and levels beyond K-12.

Adoption

Adoption was discussed in 10 of 12 discussion groups. Adoption appeared as a mid-range theme in both Los Angeles and Oklahoma City and a high-range theme in Bangor. All three sites stressed the need to emphasize the benefits of online participation to non-users. In Los Angeles, participants saw a lack of motivation as a greater barrier to adoption than affordability. Attendees at the Oklahoma City forum believed that non-users experience a lot of intimidation, frustration, and stress. Bangor participants defined the lack relevant online content as a barrier to adoption.

Digital Literacy

Digital literacy was discussed in 10 of 12 discussion groups. Digital literacy emerged as a mid-range theme in Los Angeles and Oklahoma City and a high-range theme in Bangor. All sites expressed the need to teach information literacy of online content. Los Angeles participants worried about how to provide digital literacy training outside of the education system. In Oklahoma City, they stressed the need for one-on-one training. In Bangor, attendees spoke of a digital literacy "continuum." The continuum begins with basic literacy skills and builds digital literacy upon those skills. Digital literacy training caters to individuals' needs and desires, too, instead of trying to get users to a uniform skill set.

Accessibility for People with Disabilities

Accessibility for people with disabilities was discussed in 10 of 12 discussion groups. It was a mid-range theme in Los Angeles and Bangor and a high-range theme in Oklahoma City. Los Angeles participants agreed that universal design would benefit all technology users. Both Los Angeles and Bangor attendees hoped to broaden the Framework principle to include the aspect of aging and senior citizens. Participants from Bangor and Oklahoma City shared that their local public access venues have limited assistive technology available. In Oklahoma City, participants felt that technology changes too rapidly for people with disabilities to remain up-to-date with technology skills.

Mid-Range Themes – Present in 6 -8 Discussion Groups

Collaboration

Collaboration was discussed in 8 of 12 discussion groups. It was a low-range theme in Los Angeles and Oklahoma City, but it appeared as a high-range theme at the Bangor forum. Both Los Angeles and Oklahoma City pushed for institutional partnerships among government, community-based organizations, and libraries. Oklahoma City participants struggled with the question, "Who is responsible for digital inclusion?" In Bangor, attendees overwhelming agreed that everyone, including government, libraries, schools, and technology providers, should share the work required to build the infrastructure to support digital inclusion.

Economic and Workforce Development

Economic and workforce development was discussed in 7 of 12 discussion groups. It was a mid-range theme in Los Angeles and Bangor, and a low-range theme in Oklahoma City. Both Los Angeles and Oklahoma City recognized that digital literacy is imperative for even applying for jobs. Los Angeles attendees expressed that a lack of digital training and access made it difficult for residents to find employment. Both Bangor and Oklahoma City found digital inclusion essential to economic growth on a community level. Bangor participants believed rural economies to be stymied due to a lack of technology, whereas people in Oklahoma City pushed for digital inclusion as a means to attract employers to their community. In Bangor, attendees recognized that older workers need digital training in order to remain competitive in their work field.

Quality of Life

Quality of life was discussed in 7 of 12 discussion groups. It appeared as a midrange theme in Oklahoma City and Bangor, and a low-range theme in Los Angeles. In Oklahoma, participants felt that digital inclusion improved the quality of life due to convenience and efficiency. They wanted to include "education" and "health" in the principle, too. Bangor participants believed that technology enriched the quality of life by providing a tool to maintain social connections and decrease isolation. In Los Angeles, participants found the principle too broad to properly define or articulate.

Consumer Education and Protection

Consumer education and protection was discussed in 7 of 12 discussion groups. It was a mid-range theme in Los Angeles but a low-range theme in both Oklahoma City and Bangor. Los Angeles attendees focused on the need to provide training on how to maintain equipment and software, so that people are able to upkeep broadband access at home. In Oklahoma City, participants were interested in consumer education about the many choices of technology equipment. All sites were concerned with security issues, especially when doing monetary transactions online, and privacy, like on social networking sites.

Emergency Services

Emergency service was discussed in 7 of 12 discussion groups. It was a mid-range theme in Los Angeles and a low-range theme in Oklahoma City and Bangor. In Los Angeles, participants found it important to have digital access during emergencies in order to get live information and keep in contact with family and friends. In Oklahoma City, attendees shared experiences of using Twitter to stay informed during emergencies. Also, they appreciated that the public library served as a reliable public access site during emergencies. Bangor participants hoped to push their local emergency service agencies to take better advantage of their current and rich broadband infrastructure.

Public Access

Public access was discussed by 7 of 12 discussion groups. It appeared as a midrange theme in Los Angeles and a low-range them in Oklahoma City and Bangor.

In Los Angeles, they were concerned with the limited hours of public access venues. Oklahoma City participants felt they had limited public access options due to the rural geography, which makes a many public access centers too far away from home. Both Bangor and Los Angeles attendees wanted to push for more outreach in order to increase awareness and use of existing public access venues.

Social Connectedness

Social connectedness was discussed in 6 of 12 discussion groups. It emerged as a mid-range theme in Los Angeles and Oklahoma City, but did not arise in Bangor discussions. Both Los Angeles and Oklahoma City participants saw social connectedness as a major incentive for digital adoption. Oklahoma City attendees believed older non-users are motivated to adopt in order to maintain ties with family.

Low-Range Themes – Present at Only One of the Three Forum Sites

The following themes only appeared in one of the three forums, but were discussed separately by more than one of the discussion groups. Some reflect distinguishing characteristics of the community and therefore might be more broadly applicable to other communities with those characteristics (e.g. other rural, urban, or aging communities).

Language Accessibility

This was a high-range theme in Los Angeles, which has a population with diverse language skills. Though not specifically addressed in the Proposed Framework principles, participants in the Los Angeles discussion groups emphasized the need to address diverse languages as a means to foster digital inclusion, particularly in the area of providing content. Participants advocated for more digital content available in multiple languages, and better facilitation of the use of languages other than English, especially for parents to communicate with the school system and patrons using public access technology.

This focus on accessibility for people speaking languages other than English is reflective of the diversity and large immigrant population in Los Angeles, which may not be the case in other locations. However, it may be a theme that would emerge in other urban centers with large immigrant populations, and is therefore important to note in the overall evaluation of the framework.

Smart Phones

Smart phones appeared as a mid-range theme in Los Angeles. Attendees discussed smart phones as a possible tool to promote digital inclusion. They recognized that a lot of families choose to own smart phones over home computers. Tablet computers, like the iPad, could be more commonly used tools for public access too. This focus in the Los Angeles community forum, and also other research that show the use of devices other than personal computers is growing, suggests that the Proposed Framework should be reviewed to ensure that it is device-agnostic where possible and/or that it needs to reflect the use of a variety of devices within certain goals.

Capacity Building and Investment

Capacity building and investment was a mid-range theme in Bangor. There was broad acknowledgment by participants of the state's current digital inclusion initiatives, as well as several pointed comments about capacity building in three of the discussion groups. While the participants felt that Maine was ahead of other U.S. states because of its investment in technology infrastructure (e.g. the three-ring binder project) they were forced to look outside of their borders for new ideas and service providers. Participants recognized that thoughtful investments had facilitated access to many rural areas in Maine, yet felt many people were still left behind. They began to consider where funding would come from to fill in these gaps. This theme may be reflective of an issue that may be encountered in other rural or geographically dispersed communities which needs to be more directly addressed in the Final Framework.

Senior Populations and Generational Gaps

Senior populations and generational gaps emerged as a mid-range theme in Bangor. Participants recognized Maine's senior population as a challenge to adoption and digital literacy. While some felt that seniors were simply unwilling to learn new technology, others disagreed. Many participants felt that technology education targeted to seniors was needed in Maine, and it was suggested that the younger and older generations be brought together for teaching and learning experiences. This theme was somewhat distinct from the higher level theme in which some of the discussion groups identified aging and diminishing capabilities as an aspect of the accommodation of people with disabilities.

Civic Engagement

Civic engagement was a low-range theme in Los Angeles. Participants supported use of digital technologies as a means to connect people and government institutions. One woman shared that she wanted to connect to more people in her own neighborhood. Another woman spoke of her daughter's involvement in the Libyan revolution and how her daughter uses digital technology to connect her work with people all over the world. Civic engagement, in Los Angeles, was seen as distinct from eGovernment and may reflect the need to better distinguish the two aspects of this targeted principle area in the Final Framework.

Non-Use

Non-use of digital technologies was a low-range theme in Los Angeles. As touched on in the adoption discussions, participants supported digital inclusion, but still supported people who choose to be non-adopters. They felt that digital resources should augment current options, not replace them. This theme may have also been indirectly reflected in other discussion groups and may reflect a need to take into account that for some, digital inclusion is not a desired, possible, or perhaps even necessary, condition and that it may be necessary to continue to accommodate these individuals, especially in eGovernment activities.

Health Care

Health care appeared as a low-range theme in Bangor. Participants were worried about the lack of technology infrastructure in health care facilities. They wondered if the main barrier to availability and access was policy and not cost or relevance. It was noted that mental health information, both about and for patients, is walled off from other health information. Participants felt that health care was one sector where government needed to play a driving role in digital inclusion, mostly because of its role in creating and enforcing regulatory restrictions which may affect using technology in health care. For example, HIPPA may create barriers to fully utilizing online medical records to improve access to health care information for consumers.

Right or Privilege?

Participants in two Oklahoma City focus groups wondered whether or not Internet access should be considered a right or a privilege. One person agreed that the Internet is a necessity for some people, but perhaps not for all people. Another person believed Internet access supports freedom of speech. This difference in perception led into discussions about who is responsible for enacting changes that would lead to greater digital inclusiveness. Many of the participants expressed ambivalence as to whether access to technology is an individual or group responsibility. This was a marked difference in attitude from the other locations, and in particular from Los Angeles. It may reflect a difference in the political orientation of the locations, particularly the common philosophy of the role of government. Since this political philosophy and questions about the role of government are not uncommon in many regions of the country, the Final Framework probably needs to be sensitive to, and prepared to answer, such attitudes by clarifying that its intent is to coordinate collaborative action within a community, not provoke a mandate of government intervention.

Appendix 6. Overall Analysis of the Leadership Forums

Participants in the leadership forums were given the opportunity to discuss any aspect of the framework. Overall, leaders in all of the forums affirmed that the framework was helpful in identifying issues of digital inclusion and thought that a national framework based on leading thought and practices would be valuable to individual communities.

- The framework was described as a useful and helpful tool to shape discussions, kick start digital inclusion efforts, establish guidelines, and set policy. The potential of the framework to engage and assist rural communities was also underlined.

- A concern raised in all three leadership forums was the usability of the framework in communicating with key stakeholders and local government leaders. The forum participants emphasized the need to clearly outline the relevance and benefits of information and digital technologies in order to promote the wider and greater adoption of digital technology by residents.
- The promotion of wider access and greater training in support of advances in eGovernment was another common theme in all three forums. Participants noted that access is not universal and services are not accessible to all. Training is needs to be provided to local government staff and residents in order to successfully implement, oversee, and complete eGovernment services. The economic realities of many local governments and non-profits are a barrier to adequate training.

Implementation

Clearly explaining the benefits of the digital inclusion through the framework was a challenge identified in all three leadership forums. The leaders stressed the need to communicate the benefit of digital inclusion in a way that is meaningful for all groups and outline a path to digital inclusion. Training and marketing materials need to simplify the framework for the different audiences that may use it.

General feedback on implementation was consistent across the three leadership forums and revolved around the need for the framework to address the following:

- Specific needs and circumstances in an individual community;
- The ability to dynamic and adaptable to the ever-changing nature of technology;
- Defined incentives to participate in implementation;
- The importance of information sharing across sectors;
- Recognition that certain facets of the framework, like infrastructure, are beyond a community's direct control; and
- The impact, both positive and negative, results could play in funding.

Specific Implementation Options

A suite of implementation options was presented at the leadership forums.

- Self-assessment: Recognizes self-assessed achievement. Example outlined: National Wildlife Foundation.
- Guidelines: Self-assessment that establishes a basic level of competence based on benchmarks. Examples outlined: Human Rights Campaign All Children-All Families and ILO Mainstreaming Employment and Decent Work.
- Certification: Self-assessment, outside assessment and recommendations, and a commitment to self-defined tactical activities. Example discussed: Audubon Sustainable Communities Program.
- Accreditation: Self-assessment, outside assessment (third party) and recommendations, compliance to prescriptive standards, and a commitment to self-defined tactical activities. Example discussed: New England Association of Schools and Colleges.
- Report card: Report card that ranks participants and is based on reliable, comprehensive, and comparable statistical data. Example discussed: Humanities Indicators Project.

Throughout the leadership forums, self-assessment emerged as the most viable implementation option. There appeared to be no support for a rigorous accreditation program or other approach that would be seen as expensive and/or complicated. The monetary cost and staff time typically required for such processes were seen as unrealistic commitments for digital inclusion efforts. It was also noted that the value of the accreditation was largely determined by the awarding agency/organization/entity. Allocating resources to staff training rather than the accreditation process was articulated as an option that would provide longer term benefits to the community.

The feedback was mixed, however, on the need for a recognition program for communities who implemented the framework. Participants did think that collecting and sharing effective practices and stories would be helpful in demonstrating the benefits of digital inclusion and value proposition of participation.

Forum participants thought that a national score card or index ranking program could be potentially helpful, but were concerned about the validity of the data and how the ranking might be used in negative ways.

In summary, the most important feedback on implementation was the importance of effective marketing materials and the development of tools that can be self-administered at relatively low cost.

APPENDIX 7. ONLINE SURVEY

In order to gather additional feedback about the digital inclusion framework, the IMLS posted an online survey and distributed a press release to the Digital Inclusion Working Group, the Digital Inclusion Network, participants in the community and leadership forums, and other organizations that have previously expressed an interest in the Building Digitally Inclusive Communities initiative. ICMA and the UW team also distributed invitations to their constituents.

The survey contained questions about the *Building Digitally Inclusive Communities* materials, as well as attitudinal questions about whether the framework will support digital inclusion activities. An open-ended question was also included to solicit suggestions for other improvements to the digital inclusion effort. The survey was approved by the Office of Management and Budget under a Generic Clearance for the Collection of Qualitative Feedback on Agency Service Delivery.

Survey Results

A total of 218 individual responded to the online survey. The majority of respondents reported living in a city or suburban area, while 27% reside in a rural area or remote town. 42% of respondents work for a public or school library, 35% work for a government agency, 7% work for a K-20 educational institution, 5% work for a nonprofit or community-based organization, and 2% work in the private sector. 6% were employed in another capacity including in an academic library, as students, or as volunteers. An additional 3% said they had a personal interest in digital inclusion.

In a forced choice exercise, respondents were asked to choose the three most important digital inclusion principles to them personally (Figure 3). The foundational principles of availability & affordability and public access were selected most frequently, followed by the targeted principle of education.

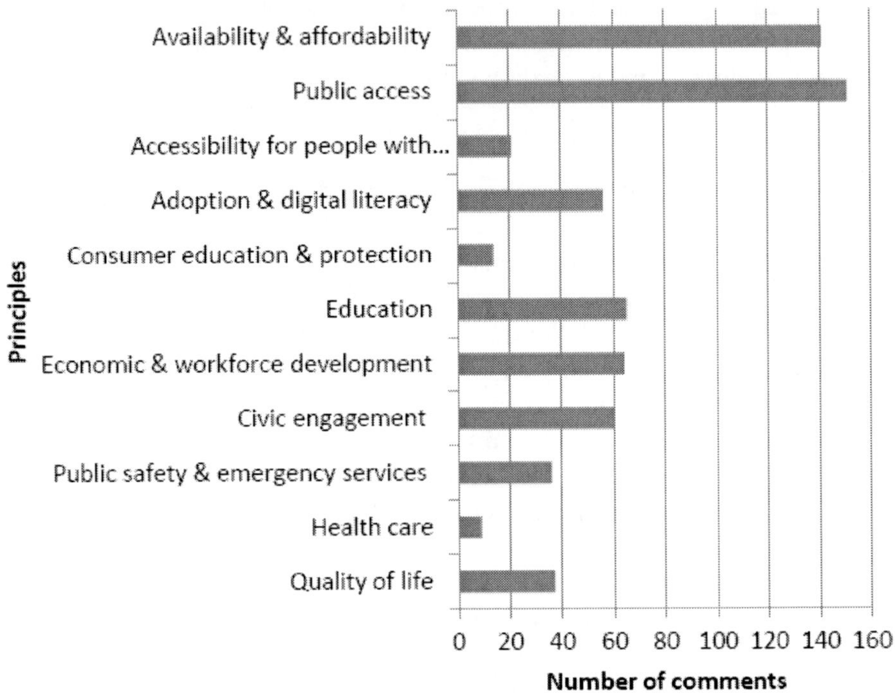

Figure 3. Ranking of Framework principles in web survey.

Respondents were asked a series of specific questions about the Framework for Digitally Inclusive Communities and the materials available to support digital inclusion activities. 72% of respondents strongly or slightly agreed that the framework would support collaboration between individuals, businesses, community-based organizations, and government in their communities. Further, 70% of respondents said their community would be likely to use some of the principles and goals outlined in the Framework for Digitally Inclusive Communities to improve digital inclusion.

Finally, each respondent was asked to select from a list of predetermined publications or tools to indicate which would helpful to their community when implementing a digital inclusion strategy. Of these, web-based training, a press kit, brochures, and a community forum were selected most (Figure 4). Of the 22 respondents who selected "Other" and provided written recommendations, many indicated that social media outlets such as Facebook could be used to distribute information about digital inclusion. Additional suggestions included best practice documents and creating an online community of practice for digital inclusion.

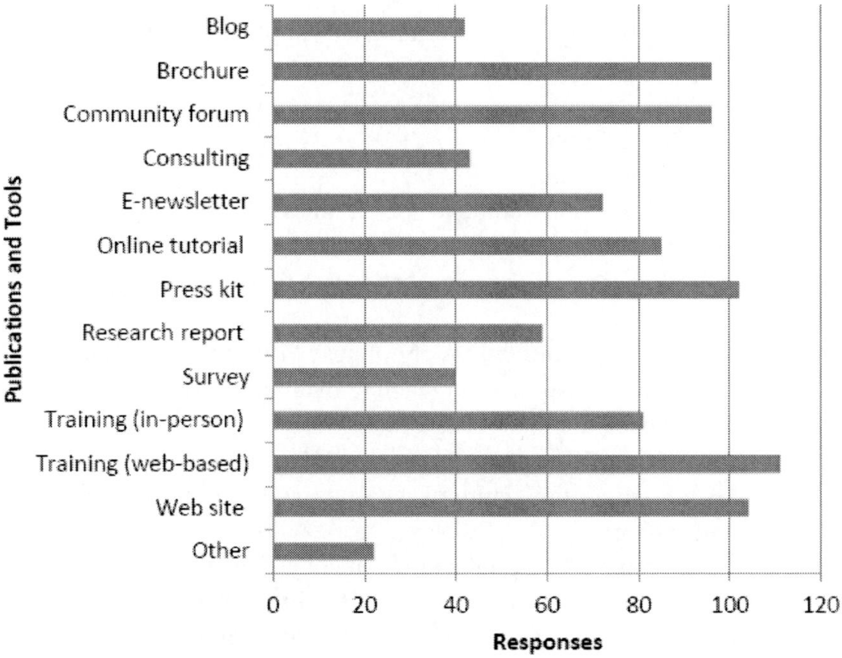

Figure 4. Perceptions of helpfulness of publications and tools in implementing a digital inclusion strategy.

Analysis of Open-Ended Comments

Content analysis of comments left by respondents revealed several themes:

- Materials and training are needed to support digital inclusion and the adoption of the principles and goals. More specific steps towards implementation need to be included in support materials for Framework implementation. Specific suggestions included:
    - Include examples of successful implementation in materials
    - Offer open-source or online tools for support
    - 'Getting Started' guide is not as helpful or practical as it could be
    - Add a section about myths and misinformation
    - Collaboration seems very easy the way the framework lays it out, but it is not; the framework should be more specific about what a "community" means and who is accountable

    - Offer training to librarians or periodic events to support implementation
    - Offer presentations and visuals and facilitator guides for implementation
    - Use social media
- The structure and content of framework could be improved. Suggestions for improvement included:
    - Clarify the difference between foundational and targeted principles
    - Emphasize affordability, or collapse affordability & availability
    - Change accessibility language to make it less disability-centric
    - Scale back the model to be more focused
    - Add more about user privacy and empowerment
    - Addressing the cultural reasons people have not adopted technology
    - Add creating and publishing content as a goal
- The relevance and future efficacy of the Framework is not clear, nor is it clear who is responsible for implementing digital inclusion activities. Specifically:
- Funding is not addressed in Framework and it's not clear whether it is a call for commitment of public funds
- The private sector is not figured prominently enough in the conversation
- IMLS needs to take a greater leadership role in effort and seek greater visibility for the effort
- The Framework needs coordinated support at the state government level
- Libraries may have a difficult time supporting the Framework as they are already strained

REFERENCES

Federal Communication Commission (FCC). (2010). *Connecting America: The National Broadband Plan.* Washington DC: Federal Communications Commission. Retrieved from http://download.broadband.gov/plan/national-broadband-plan.pdf

Mace, R., Hardie, G, Place, J. (1991). Accessible environments: Toward universal design. In W.E. Preiser, JC. Vischer, & E.T. White (Eds.),

Design Intervention: Toward a More Humane Architecture. New York, NY: Van Nostrand Reinhold.

BIBLIOGRAPHY

This bibliography is divided into four topic areas: assessment models; indicator systems; local and international policy initiatives on digital inclusion; and reasons for non-adoption.

Assessment Models

American Library Association. (2003) *Principles for the networked world.* Retrieved from http://www.ala.org/ala/aboutala/offices/oitp/ publications /issuebriefs/principles/principles.pdf

Beetham, D., Carvalho, E., Landham, T., and Weir, S. (2008). *Assessing the quality of democracy.* Stockholm, Sweden: International Institute for Democracy and Electoral Assistance. Retrieved from http://www.dpwg-lgd.org/cms/upload/pdf/quality_of_democracy_a_practical_guide_idea. pdf

Bryne Potter, A. (2006). Zones of silence: A framework beyond the digital divide. *First Monday, 11*(5).

Grim, T. (2008). Foresight maturity model [unpublished slide presentation]. Retrieved from http://www.foresightmaturitymodel.com

Harvey, M. (n.d.). *Discussion of the Environmental Literacy Council Framework.* Retrieved from http://www.enviroliteracy.org/ article.php /449.html

Human Rights Campaign Foundation, All Children-All Families Initiative. (2010). *Promising practices in adoption and foster care: A comprehensive guide to policies and practices that welcome, affirm and support lesbian, gay, bisexual and transgender foster and adoptive parents* (3rd ed.). Washington, DC. Retrieved from http://www.hrc.org/documents/ HRC_Foundation_Promising_Practices_Guide_3rd_Edition_Color.pdf

Institute of Museum and Library Services. (n.d.). *Museums, libraries, and 21st century skills self-assessment tool.* Retrieved from http://imls21stcentury skills.org/

National Information Standards Organization. (2007). *A framework of guidance for building good digital collections* (3rd ed.). Retrieved from http://framework.niso.org/

U.S. Green Building Council. (n.d.). *LEED certification rating system.* Retrieved from http://www.usgbc.org/DisplayPage.aspx?CMSPageID=51

_______. (n.d.). *LEED rating system development.* Retrieved from http://www.usgbc.org/DisplayPage.aspx?CMSPageID=2360

Wynne, M. E., & Cooper, L. F. (2007). *Power Up: The campaign for digital inclusion — Digital inclusion imperatives offer municipalities new social and economic opportunities.* Bellingham WA: Microsoft Corporation.

Wynne, M. E., Perry, K. A., Cooper, L.F. (2008). *Innovating for inclusion: A digital inclusion guide for those leading the way.* Bellingham, WA: Microsoft Corporation.

Indicator Systems

Abraham, K. (2006). Nonresponse in the American Time Use Survey: Who is missing from the data and How much does it matter? *Public Opinion Quarterly, 70*(5): 676–703.

ACT Rochester. (n.d.). *ACT Rochester: Community Indicators for the Greater Rochester Area.* Retrieved from http://actrochester.org/

Advancing Our Community. (n.d.). *Sitka community indicators.* Retrieved from http://sitkaaoc.org/data/indicators.html

Alliance for Building Community. (n.d.). *Community building resources.* Retrieved from http://www.allianceforbuildingcommunity.org/

Aloha United Way. (2005). *Quality of life in Hawaii.* Honolulu HI: University of Hawaii at Manoa, College of Tropical Agriculture and Human Resources, Center on the Family. Retrieved from http://uhfamily.hawaii.edu/ publications/publications_archive.asp

Annie E. Casey Foundation. (2010). *2010 Kids Count data book: State profiles of child well-being.* Retrieved from http://datacenter.kidscount.org/DataBook/2010/Default.aspx

Applied Survey Research. (2010). *Santa Cruz County Community Assessment Project Year 16, 2010.* Retrieved from http://www.santacruzcountycap.org/ pdfs/CAP%2016/Complete%20CAP%20Report%20-%20Year%2016.pdf

Asheville Hub. (n.d.). *Asheville Hub.* Retrieved from http://www.ashevillehub.com/

Austin City Connection. (n.d.). *Austin sustainable communities initiative.* Retrieved from http://www.norfolk.gov/mayor/trees/documents/green%20team/research%20and%20examples/city%20of%20austin%20-%20sustainable%20communities%20initiative.htm

Australian Unity. (2010). *Australian unity wellbeing index.* Retrieved from http://www.australianunitycorporate.com.au/community default.aspx

Baltimore Neighborhood Indicators Alliance and Jacob France Institute. (n.d.). *Neighborhood data.* Retrieved from http://www.bniajfi.org/

Boone County. (2009). *Boone County indicators.* Retrieved from http://booneindicators.org/

Boston Foundation, The. *The Boston indicators project.* Retrieved from http://www.bostonindicators.org/Indicators2008/

Burlington Legacy Project. (2000). *Burlington legacy action plan.* Burlington, VT: City of Burlington.

Calgary Foundation, The. (2009). *Calgary's vital signs: 2009 citizens' report card.* Retrieved from http://www.thecalgaryfoundation.org/documents/09VitalSignsFinal_001.pdf

Canadian Council on Learning. (2010). Composite learning index. Retrieved from http://www.cli-ica.ca/en.aspx

Canmore Community Monitoring Program. (2008). Community indicators. *Biosphere Institute of the Bow Valley.* Retrieved from http://www.biosphereinstitute.org/?q=r-com-indicators

Carver County. (2006). *Quality of life indicators.* Chaska, MN: Carver County Public Health Division, Land and Water Services Division, and Land Management Department.

Center for Houston's Future. (2009). *Counting on quality of place: Air quality, parks & trails, trees.* Houston, TX.

Central Texas Sustainability Indicators Project. (2009). *2009 indicators report.* Austin, TX. Retrieved from http://www.centex-indicators.org/

Chamard, S., & Parker, K. (2009). *The Matanuska-Susitna Borough community survey, 2009: A sourcebook of community attitudes.* Anchorage, AK: Justice Center, University of Alaska Anchorage.

Champlain Initiative. (n.d.). *The changing face of Chittenden County: A multi-sector partnership embracing our changing community.* Retrieved from http://www.champlaininitiative.com/home

Chapin Hall. (n.d.). *Chapin Hall at the University of Chicago.* Retrieved from http://www.chapinhall.org/

Chester County Planning Commission. (2010). *Chester County landscapes progress report 2010, landscape index.* Chester County, PA.

Children First for Oregon. (2009). *2009 status of Oregon's children: Keeping children safe – Improving Oregon's child welfare system.* Portland, OR.

Children's Services Council of Broward County. (n.d.). *Children's strategic plan.* Lauderhill, FL. Retrieved from http://www2.cscbroward.org/ Strategic.aspx

Children's Workforce Development Council. (2010). *The common assessment framework for children and young people: A guide for managers.* Leeds, UK. Retrieved from http://publications.cwdcouncil.org.uk/ product_info. php?products_id=199

City of Baltimore. (n.d.). *Baltimore CitiStat.* Retrieved from http://www. baltimorecity.gov/ Government/AgenciesDepartments/ CitiStat.aspx

City of Calgary. (2006). *2006 state of the environment report* (3rd ed.). Calgary, Canada.

City of Calgary. (2008). *Towards a sustainable city: The case for change.* Calgary, Canada.

City of Calgary. (2009). *Plan it Calgary.* Retrieved from http://www. calgary.ca/planit

City of Greensboro. (n.d.). *GreensbOrometer.* Retrieved from http://www. greensboro-nc.gov/departments/hcd/planning/Greensb Orometer.htm

City of Minneapolis. (n.d.). *Results Minneapolis.* Retrieved from http://www.ci.minneapolis.mn.us/results/

City of Northampton, Office of Planning and Development. (2008). *Sustainable Northampton: Comprehensive plan, January 2008.* Northampton, MA.

City of Pasadena, Public Health Department. (2002). *Quality of life index: 2002 index.* Pasadena, CA.

City of Rocklin. (2010). *Community report card 2009–2010.* Rocklin, CA.

City of Santa Monica. (2010). *Sustainable Santa Monica: Sustainable city report card 2010.* Retrieved at http://www.smgov.net/uploadedFiles/ Departments/OSE/Categories/Sustainability/Sustainable_City_Report_C ard_2010.pdf

City of Surrey. (2008). *Sustainability charter: A commitment to sustainability.* Surrey, British Columbia.

City of Vancouver. (n.d.). *Strategic plan/strategic commitments.* Retrieved from http://www.cityofvancouver.us/StrategicPlan.asp?year=2008& item ID =52317

City of Winnepeg. (n.d.). *Peg.* Retrieved from http://www.mypeg.ca/

Communities Count. (n.d.). *Social and health indicators across King County.* Retrieved from http://www.communitiescount.org/index.php

Community Action Network. (n.d.). *Community Action Network.* Retrieved from http://www.caction.org/

Community Agenda. (2010). *Community agenda snapshot: Volusia and Flagler counties, June 2010.* Retrieved from http://www.agendavf.org/ 2010%20Snapshot%20FINAL.pdf

Community Foundation of Ottawa. (2010). *Ottawa's vital signs: The city's annual checkup.* Ottawa, Canada.

Community Foundation, The. (2009). *Boulder County trends report 2009.* Retrieved from http://www.commfound.org/nonprof/ civicforum/index. php

Community Indicators Consortium. (n.d.). *Indicators projects*. Retrieved from http://www.communityindicators.net/efforts

Community Indicators Initiative of Spokane. (n.d.). *Spokane community indicators.* Retrieved from http://www.communityindicators.ewu.edu/

Community Indicators Victoria. (n.d.). *Community indicators: Victoria wellbeing reports.* Retrieved from http://www.communityindicators.net.au / wellbeing_reports

Community Quality Improvement. (2009). *Beyond compassion: Measuring the external costs of poverty in Sault Ste. Marie.* Sault Ste. Marie, Canada: Social Services Administration Board.

Community Resources Council. (n.d.). *MAACLink.* Retrieved from http://www.crcnet.org/

Community Vision. (2010). *Osceola County report card: Focusing on community connections* (Vol. 7). Retrieved from http://www.community vision.org/ Documents/ReportCard%202010.pdf

Coordinating Council, The. (2009). *Calhoun County community report card.* Retrieved from http://www.tcccalhoun.org/CRC/2009%20Community% 20Report%20Card%20and%20Trees.pdf

Costa, S., Palaniappan, M., Wong, A.K., Hays, J., Landeiro, C., & Rongerude, J. (2002). *Neighborhood knowledge for change: The West Oakland environmental indicators project.* Oakland, CA: Pacific Institute for Studies in Development, Environment, and Security.

Council on Virginia's Future. (2010). *The Virginia report.* Retrieved from http://www.future.virginia.gov/docs/VirginiaReport/2010_Virginia_%20R eport(Web)_compressed.pdf

County of Summit. (2009). *Summit 2010 priority indicators progress report.* Retrieved from http://www.healthysummit.org/qol/pdf/ final%20draft%20 2009%20data%20tracking%20report%200216.pdf

Dallas Foundation and Foundation for Community Empowerment. (n.d.) *Dallas indicators.* Retrieved fromhttp://www.dallasindicators.com/

Delaware Valley Regional Planning Commission. (n.d.). *Regional indicators.* Retrieved from http://www.dvrpc.org/LongRangePlan/Regional Indicators/

Department for Environment, Food, and Rural Affairs. (2010). *Measuring progress: Sustainable development indicators 2010.* London, United Kingdom.

Department of Community Affairs. (n.d.). *Georgia community indicators.* Retrieved from http://www.dca.state

Department of Employment and Economic Development. (n.d.). *Minnesota 3D web mapping.* Retrieved from http://map.deed.state

Eastern Washington University, Institute for Public Policy and Economic Analysis. (n.d.). *Northeast Washington trends.* Retrieved from http://www.northeastwashingtontrends.ewu.edu/hiSpeed/index.cfm

Eastern Washington University. (n.d.). *Chelan Douglas trends.* Spokane, WA: Eastern Washington University Institute for Public Policy and Economic Analysis & the Center for Digital Media Design and Development. Retrieved from http://www.chelandouglastrends.org/

Eastern Washington University. (n.d.). *Grant County trends.* Retrieved from http://www.grantcountytrends.ewu.edu/

Environment Canada. (n.d.). *Measuring sustainability: Canadian environmental sustainability indicators.* Retrieved from http://www.ec.gc.ca/indicateursindicators/

Equinox Center. (2010). *The San Diego regional quality of life dashboard.* San Diego, CA.

Family League of Baltimore City. (2007). *How are we doing? Indicator data on child and family well being.* Baltimore, MD: Baltimore City Data Collaborative.

Fond du Lac Area United Way. (2007). *2007 local indicators for excellence: LiF.E. in Fond du Lac County.* Fond du Lac, WI.

Foundation of Greater Montreal. (2010). *Greater Montreal's vital signs: Our region's annual check-up.* Montreal, Canada.

Franklin Community Health Network. (n.d.). *Healthy community coalition.* Retrieved from http://www.fchn.org/hcc

Fransman, M. (2006). *Global broadband battles: Why the U.S. and Europe lag while Asia leads.* Stanford, CA: Stanford Business Books.

Fraser Basin Council. (2010). *Working together in the lower mainland: Sustainability snapshot 2010. A report of the Fraser Basin Council for organizations and residents of: Fraser Valley, Metro Vancouver, Squamish-Lillooet.* Vancouver, Canada.

Government of Alberta. (2010). *Measuring up: Progress report on the Government of Alberta business plan.* Edmonton, Canada.

Great Valley Center. (n.d.). *Indicators.* Retrieved from http://www.greatvalley.org/indicators/

Greater New Orleans Community Data Center. (2010). *The New Orleans index at five: Measuring greater New Orleans' progress toward prosperity.* New Orleans, LA: Brookings Metropolitan Policy Program & Greater New Orleans Community Data Center.

Hennepin County Minnesota. (2006). *Community indicators report.* Retrieved from http://www.co.hennepin.mn.us/portal/site/HennepinUS/ menuitem.b1ab75471750e40fa01dfb47ccf06498/?vgnextoid=4e32d18feb 5b4210VgnVCM10000049114689RCRD

Hillsborough County. (n.d.). *Hillsborough community atlas.* Retrieved from http://www.hillsborough.communityatlas.usf.edu/

Howard County Department of Citizen Services. (2009). *People helping people: Annual report 2009.* Retrieved from http://www.howard countymd.gov/ CitizenServices/CSDocs/dcsar9.pdf

Human Services Council, The. (2010). *Community indicators report.* Norwalk, CT.

ICMA. (n.d.). *Center for sustainable communities.* Retrieved from http://icma.org/en/icma/priorities/sustainable_communities

_______. (n.d.). *ICMA center for performance measurement.* Retrieved from http://icma.org/en/Page/107/ICMA_Center_for_Performance_Measurem ent

Indiana University Purdue University Indianapolis (IUPUI) & the Polis Center. (n.d.). *The SAVI Community Information System.* Retrieved from http://www.savi.org

Indicators Progress Commission. (2008). *Albuquerque progress report 2008.* Albuquerque, NM: City of Albuquerque. Retrieved from http://www.cabq.gov/progress/ documents/complete_2008_apr.pdf

Information Hub of South Central Connecticut, The. (n.d.). *Data haven.* Retrieved from http://www.ctdatahaven.org/

International Monetary Fund. (2003). *Data quality assessment framework (DQAF) for external debt statistics.* Retrieved from http://dsbb.imf.org /vgn/ images/pdfs/dqaf_exd_statistics.pdf

Jacksonville Community Council Inc. (2009). *Quality of life progress report for Jacksonville and Northeast Florida* (25th ed.). Retrieved from http://www.jcci.org/jcciwebsite/documents/09%20QOL%20Summary%20Document.pdf

Kaua'i Planning & Action Alliance. (2010). *Measuring what matters for Kaua'i – Community indicators report 2008.* Retrieved from http://www.kauainetwork.org/wp-content/uploads/2010/02/ indicatorsfinal report2008.pdf

Kazanjian, A., and C. J. Green. (2002). Beyond effectiveness: The evaluation of information systems using a comprehensive health technology assessment framework. *Computers in Biology and Medicine, 32*(3): 165–177.

Kentucky Youth Advocates. (n.d.). *2010 Kentucky kids count county data book.* Retrieved from http://www.kyyouth.org/documents/kya_ databook 10.pdf

Kids County North Dakota. (2010). *North Dakota kids count 2010 fact book: State, regional, and county profiles of child well-being in North Dakota.* Retrieved from http://www.ndkidscount.org/factbook/ completefactbook /NDKCFactBook_2010.pdf

Lancaster County Planning Commission. (n.d.). *Long-range planning.* Retrieved from http://www.lancastercountypa.gov/planning/cwp/browse.asp?A=2

Larimer County Department of Human Services. (n.d.). *Compass of Larimer County.* Retrieved from http://www.co.larimer.co.us/compass/

Leadership Alliance for a Vital Community. (n.d.). *A community indicators project for Herkimer and Oneida Counties.* Retrieved from http://www.communityprofiles.org/herkimeroneida/

League for Innovation in the Community College. (2004). *An assessment framework for the community college: Measuring student learning and achievement as a means of demonstrating institutional effectiveness.* Retrieved from www.league.org/publication/whitepapers/files/0804.pdf

Leibniz Institute for Social Sciences. (n.d.). *German system of social indicators.* Retrieved from http://www.gesis.org/en/services/

Lucille Packard Foundation. (n.d.). *Kidsdata.* Retrieved from http://www.kidsdata.org/

Maine Center for Economic Policy. (2009). *Ideas for shared prosperity.* Retrieved from http://www.mecep.org/

Maine Economic Growth Council. (2010). *Measures of growth in focus.* Retrieved from http://www.maine.com/editions/2006-02-15/images 20060202000100C.pdf

Maricopa County, Office of Management and Budget, Managing for Results. (2006). *Maricopa County annual report of community indicators,* 2006. Retrieved from http://www.maricopa.gov/mfr/pdf/Com_Indicators.pdf

Massarand, M. G. (2010). 2010 *Analysis of impediments to fair housing and community resources in Idaho: An assessment of perceptions across multiple stakeholder groups.* Boise, ID: Interaction Consulting International, Inc.

Mid-American Regional Council. (2001*). Metro outlook: Measuring the progress of Metropolitan Kansas City.* Retrieved from http://www. marc.org/ Metro%20Outlook.pdf

Ministry for the Environment. (2008). *Environment New Zealand 2007.* Retrieved from http://www.mfe.govt.nz/publications/ser/enz07-dec07/

Minnesota Compass. (n.d.). *Minnesota compass.* Retrieved from http://www.mncompass.org/

Minnesota Department of Administration. (2002). *Minnesota milestones: Measures that matter*. Retrieved from http://www.mnplan.state

Missoula County. (n.d.). *Missoula measures.* Retrieved from http://www.co. missoula.mt.us/measures/

Morrison Institute for Public Policy. (n.d.). *Arizona indicators.* Retrieved from http://arizonaindicators.org/

Murley, J. F. (2008). *Preserving paradise: SoFlo's call to action.* Boca Raton, FL: Florida Atlantic University, Center for Urban & Environmental Solutions.

Murphey, D. (2006). *Vermont well-being: A social indicators sourcebook.* Waterbury, VT: Vermont Agency of Human Services, Planning Division.

National Association of Counties. (2010*). 2010 county sustainability strategies.* Retrieved from http://www.naco.org/newsroom/latest /Documents/ 2010%20Sustainability%20Strategies%20Publication.pdf

National Association of Local Government Environmental Professionals. (2003). *Smart growth for clean water: Helping communities address the water quality impacts of sprawl.* Retrieved from http://www.nalgep.org/ publications/PublicationsDetail.cfm?LinkAdvID=42157

New York State Council on Children and Families. (2005). *New York State touchstones, kids count data book.* Retrieved from http://www. ccf.state2009TouchstonesDB.pdf

New Yorkers for Parks. (2009). *New Yorkers for parks 2009 annual report.* Retrieved from http://www.ny4p.org/index.php

North Yorkshire County Council. (n.d.). *Neighbourhood profiles.* Retrieved from http://www.northyorks.gov.uk/index.aspx?articleid=9701

Orange County. (2010). *Orange County 2010 community indicators.* Retrieved from http://www.ocgov.com/vgnfiles/ocgov/CEO/Docs/ 2010%20 Community%20Indicators.pdf

Oregon Progress Board. (2009). *Highlights: 2009 benchmark report to the people of Oregon.* Retrieved from http://www.oregon.gov /DAS/OPB/ docs/2009Report/2009_Benchmark_Highlights.pdf

Oregon State University. (n.d.). *Rural studies program.* Retrieved from http://ruralstudies.oregonstate.edu/

Pikes Peak United Way. (2010). *2010 quality of life report.* Retrieved from http://pikespeakqualityoflife.org/2010/09/19/2010-pikes-peak-quality-of-life-indicators-report-now-available/

Pinellas County. (n.d.). *Pinellea County community indicators.* Retrieved from http://www.pinellasindicators.org/

Piton Foundation, The. (n.d.). *Community facts.* Retrieved from http://www.piton.org/CommunityFacts.

Planning Council, The. (2008). *2008 Regional report card, South Hampton Roads, VA.* Retrieved from http://www.theplanningcouncil.org/ ReportCard08.pdf

Prosperity Pensacola. (2006). *Escambia County indicator report.* Retrieved from http://uwf.edu/haas/pdfFiles/Indicator_Report_revised.pdf

Province of Manitoba. (2009). *Provincial sustainability report for Manitoba.* Retrieved from http://www.gov.mb.ca/conservation sustainabilty_ report_2009.pdf

Public Policy Institute of Marion County. (2005). *2005 quality of life progress report.* Marion County, FL.

Puget Sound Partnership. (n.d.). *Puget Sound dashboard of ecosystem indicators.* Retrieved from http://www.psp.wa.gov/pm_dashboard.php

Rauch Foundation. (n.d.). *Long Island index.* Retrieved from http://www.longislandindex.org/

Rochdale Borough Council. (2010). *Empowering local communities to make better decisions: Case study.* Retrieved from http://www.instantatlas.com/ downloads/Rochdale_Council_CaseStudy.pdf

Royal Children's Hospital, Melbourne. (n.d.). *Australian early development index: Building better communities for children.* Retrieved from http://maps.aedi.org.au/

San Antonio Metropolitan Health District. (n.d.). *Metro health: San Antonio's public health team.* http://www.sanantonio.gov/health/healthprofiles-main.html

San Joaquin County. (2008). *Healthier San Joaquin County community assessment*. Retrieved from http://www.healthiersanjoaquin.org/

Sarasota County Openly Plans for Excellence. (n.d.). *SCOPE: Connecting and inspiring citizens to create a better community.* Retrieved from http://www.scopexcel.org/default.asp

Security Privacy and Critical Infrastructure Committee. (2000). *Federal information technology security assessment framework.* Gaithersburg, MD: National Institute of Standards and Technology.

Sightline Institute. (n.d.). *Cascadia scorecard.* Retrieved from http://scorecard.sightline.org/

Society for the Protection of New Hampshire Forests. (2005). *New Hampshire's changing landscape.* Retrieved from http://www.forest society. org / research/research-projects.asp

South Australia Strategic Plan. (n.d.). *Plan for the best.* Retrieved from http://www.saplan.org.au/

Southern California Association of Governments. (2007). *The state of the region 2007: Measuring regional progress.* Retrieved from http://www. scag.ca.gov/sotr/index.htm

Southern Illinois Healthcare. (n.d.). *Healthy communities coalition.* Retrieved from http://www.sih.net/home.nsf/content/ healthycommunitiescoalition3

Spartanburg County Foundation, The. (2005). *Community indicators VI: Strategic Spartanburg.* Retrieved from http://www.strategicspartanburg. org/ documents.php

State of Connecticut. (2009). *Connecticut commission on children, 2008–2009 annual report.* Retrieved from www.cga.ct.gov/coc/PDFs/annualreports/ 2009_annual_rcport.pdf

State of Guernsey Policy Council. (2009). *Sustainable Guernsey: Monitoring fiscal and economic, social, and environmental trends.* Retrieved from http://www.gov.gg/ccm/cms-service/download/asset/ ?asset_id=13564678

State of Iowa. (n.d.). *Results Iowa: Accountability for Iowa.* Retrieved from http://www.resultsiowa.org/index.html

Sudbury Community Foundation. (2010). *Greater Sudbury's vital signs 2010.* Retrieved from http://vitalsignscanada.ca/local/sudbury/pdf/ Sudbury_ VitalSigns%20Report_2010.pdf

Sustainability Indicators. (n.d.). *Sustainability indicators for Sunnybank, Brisbane, Queensland, Australia.* Retrieved from http:// www.sunnybank.info

Sustain Cape Cod. (n.d.). *Indicators.* Retrieved from http://www.sustain capecod.org/indicators.

Sustainable Cincinnati. (2005). *Regional indicators measuring the economic, environmental, and social health of the Tri-State Metropolitan area.* Retrieved from http://www.sustainablecincinnati.org/sc_2005.pdf

Sustainable Pittsburgh. (2004). *Southwestern Pennsylvania regional indicators report.* Retrieved from http://www.sustainablepittsburgh.org/pdf/ 2004Indicators.pdf

Sustainable San Mateo County. (2009*). Indicators for a sustainable San Mateo County: Thirteenth annual report card.* Retrieved from http://www.sustain ablesanmateo.org/indicators-report/reports/2009- indicators-report/

Sustainable Seattle. (n.d). *Sustainable Seattle.* Retrieved from http:// sustainableseattle.org/

Temple University. (n.d.). *Metropolitan Philadelphia indicators project.* Retrieved from http://mpip.temple.edu/

Therivel, R., & Wood, G. (2004). Tools for strategic environmental assessment. In M. Schmidt, E. João & E. Albrecht (Eds.), *Implementing Strategic Environmental Assessment* (349–364). Berlin, Germany: Springer.

Thomas Jefferson Planning District Commission. (2009). *2008–2009 annual report.* Retrieved from http://www.tjpdc.org/newsletter/ AnnualReport_ 08_09/index.shtm

Toronto Community Foundation. (2010). *Toronto's vital signs 2010, full report.* Toronto, Canada. Retrieved from http://www.tcf.ca/vitalinitiatives/ TVS10FullReport.pdf

Truckee Meadows Tomorrow. (n.d.). *Quality of life indicators.* Retrieved from http://www.truckeemeadowstomorrow.org/.

Tuolumne County. (n.d.). T*uolumne County profile: Community indicators project.* Retrieved from http://www.tuolumnecountyprofile.org/

U.S. Department of Health and Human Services. (2010). *Healthy people.* Retrieved from http://www.healthpeople.gov

United Nations Development Programme (2010). *Human development report 2010: The real wealth of nations, pathways to human development.* New York, NY: Palgrave Macmillan.

United Way of Central Oklahoma. (2010*). Oklahoma County vital signs.* Retrieved from http://www.unitedwayokc.org/sites/default/files/ files /documents/Healthy_Citizens_Vital_Signs_-_Small.pdf

United Way of Marathon County. (2009). *FOCUS 2009–2011, LIFE in Marathon County: Local indicators for excellence.* Retrieved from http://www.co.marathon.wi.us/is/cad/pdf/LifeReportFull.pdf

United Way of Pierce County. (n.d.). *Community indicators.* Retrieved from http://indicators.uwpc.org/index2.html

United Way of the Greater Capital Region. (n.d.). *Capital Region community profile.* Retrieved from http://www.communityprofiles.org/ capitalregion /index.php

University of Anchorage Alaska. (n.d.). *Community indicators project at UAA.* Retrieved from http://justice.uaa.alaska.edu/indicators/index.html

University of Buffalo. (n.d.). *University at Buffalo Regional Institute.* Buffalo, NY:

University at Buffalo Law School. Retrieved from http://regional-institute.buffalo

University of Idaho. (n.d.). *Indicators Northwest.* Retrieved from http://www.indicatorsnorthwest.org/

University of Pennsylvania, Department of City and Regional Planning. (2010). *Connecting for global competitiveness: Florida's super region.* Retrieved from http://www.reconnectingamerica.org/assets cfgcflsuperregion.pdf

University of Pittsburgh, Center for Social and Urban Research. (n.d.). *Pittsburgh today: Key indicators for understanding our region.* Retrieved from http://www.pittsburghtoday.org/home.html

University of South Carolina. (n.d.). *South Carolina indicators project.* Retrieved from http://www.ipspr.sc.edu/scip/

University of South Dakota. (2009). *The 2009 South Dakota kids count factbook.* Retrieved from http://www.usd.edu/business/south-dakota-kids-count/upload/2009Factbook.pdf

Vancouver Foundation. (2010). *Vital signs for Metro Vancouver.* Retrieved from http://www.vancouverfoundationvitalsigns.ca/

Victoria Foundation. (2009). *Victoria's vital signs.* Retrieved from http://intraspec.ca/vital_signs_2009.pdf

Vision for Children at Risk. (2009). *Children of Metropolitan St.Louis: Report to the community.* Retrieved from http://visionforchildren.org/aboutus/ publications

Washington County. (2009). *2009 Performance measures and indicators report.* Retrieved from http://www.co.washington.mn.us/_asset/znv7b5 /ADM2010-WCPerfReportFINAL2009.pdf

Welborn Baptist Foundation. (2006). *Leading community health indicators, 2006.* Retrieved from http://www.welbornfdn.org/2006%20Community% 20 Indicators%20Report.pdf

West Midlands Regional Observatory. (2010). *State of the West Midlands 2010.* Retrieved from http://www.wmro.org/stateofthewestmidlands2010/
Whistler 2020 (2009). *Whistler 2020: Moving toward a sustainable future.* Retrieved from http://www.whistler2020.ca/whistler/site/homepage.acds?instanceid=1930792&context=1930501
World Bank. (2010). *World development indicators.* Retrieved from http://data.worldbank.org/
World Health Organization. (2008). *Primary health care: Now more than ever.* Geneva, Switzerland.
Yampa Valley Partners. (n.d.). *Community indicators.* Retrieved from http://www.yampavalleypartners.com/community
YorkCounts. (2010). *YorkCounts 2010 annual report.* Retrieved from http://www.yorkcounts.org/YorkCounts2010.pdf

Local and International Policy Initiatives on Digital Inclusion

Al-Omari, A., & Al-Omari, H. (2006). e-Government readiness assessment model. *Journal of Computer Science, 2*(11): 841–845.
Americorps. (2008). Community Technology Empowerment Project. Retrieved from http://wip.technologypower.org/
Arzberger, P., Schroeder, P., Beaulieu, A., Bowker, G., Casey, K., Laaksonen, L., Moorman, D., Uhlir, P., & Wouters, P. (2010). An international framework to promote access to data. *Science, 303*(5665): 1777–1778.
Barzilai-Nahon, K. (2006). Gaps and bits: Conceptualizing measurements for digital divide/s. *The Information Society, 22*: 269–278. Retrieved from http://uwnews.washington.edu/ni/relatedcontent/2006/October/rc_parentID27342_thisID27365.pdf
Bush, N.E., Bowen, D.J., Wooldridge, J., Ludwig, A., Meischke, H., & Robbins, R. (2004). What do we mean by Internet access? A framework for health researchers. *Preventing Chronic Disease, 1*(4).
California Emerging Technology Fund. (2008). *California ICT digital literacy assessment and curriculum framework.* Retrieved from http://www.ictliteracy.info/rf.pdf/California%20ICT%20Assessments%20and%20Curriculum%20Framework.pdf
Chung, Y.H. (2002). *Korean national policies to close the digital divide.* Retrieved from unpan1.un.org/intradoc/groups/public/documents/APCITY/ UNPAN006369.pdf

City of Chicago. (n.d.). *Digital excellence initiative.* Retrieved from http://www.cityofchicago.org/city/en/depts/doit/provdrs/dei.html

City of Minneapolis. (n.d.). *Wireless Minneapolis.* Retrieved from http://www.ci.minneapolis.mn.us/wirelessminneapolis/index.asp

City of Seattle. (2009). *Information technology access and adoption in Seattle: 2009 resident survey report.* Seattle, WA.

Distaso, W., Lupi, P., & Manenti, F.M. (2006). Platform competition and broadband uptake: Theory and empirical evidence from the European Union. *Information Economics and Policy, 18*: 87–106.

European Commission. (2008*). Digital literacy European Commission working paper and recommendations from digital literacy high-level expert group.* Retrieved from http://ec.europa.eu/information_society i2010/docs/digital_literacy/digital_literacy_review.pdf

Federal Communication Commission (FCC). (2010). *Connecting America: The National Broadband Plan.* Washington DC: Federal Communications Commission. Retrieved from http://download.broadband.gov/plan/national-broadband-plan.pdf

Fransman, M. (2006). Global broadband battles: *Why the U.S. and Europe lag while Asia leads.* Stanford, CA: Stanford Business Books.

Holland, P., Rajabifard, A., & Williamson, I. (2010). *Towards an assessment framework for spatially enabled government.* Paper presented at GSDI 12: Realising Spatially Enabled Societies, Singapore.

i2010 High Level Group. (2009). *Benchmarking digital Europe 2011–2015: A conceptual framework.* Retrieved from http://ec.europa.eu/information_society/eeurope/i2010/docs/benchmarking/benchmarking_digital_europe_2011-2015.pdf

Intelligent Community Forum. (n.d.). *Community accelerator.* Retrieved from https://www.intelligentcommunity.org/index.php?src=gendocs&ref=Accelerator&category=Services

_______. (n.d.). *Intelligent community indicators.* Retrieved from https://www.intelligentcommunity.org/index.php?submenu=Research&src=gendocs&ref=Research_Intelligent_Community_Indicators&category=Research

Kent, D., & McClure, D.P. (2009). *Digital inclusion: Bringing the rest of American online with broadband.* Luray, VA: U.S. Internet Industry Association.

Livingston, S., & Helsper, E. (2007). Gradations in digital inclusion: Children, young people and the digital divide. *New Media Society, 9*(4): 671–696.

Locke, G., & Strickling, L.E. (2010). *Plan and timetable to make available 500 Megahertz of spectrum for wireless broadband.* Washington, DC: U.S. Department of Commerce.

National Telecommunications & Information Administration. (2008). *Networked Nation: Broadband in America 2007.* Washington, DC: U. S. Department of Commerce.

Newsom, G., & and the Department of Telecommunications & Information Services. (2007). *San Francisco digital inclusion strategy.* San Francisco, CA: Department of Telecommunications & Information Services, City & County of San Francisco. Retrieved from: http://www.sfgov3.org/ftp/uploadedfiles/dtis/tech_connect/DigitalInclusionStrategy.doc

Office of Management & Budget. (2009). *Improving agency performance using information and information technology (Enterprise architecture assessment framework v3.1).* Retrieved from http://www.whitehouse.gov/sites/default/files/omb/assets/fea_docs/OM B_EA_Assessment_Framework_v3_1_June_2009.pdf

Press, L., Burkhart, G., Foster, W., Goodman, S., Wolcott, P. & Woodard, J. (1998). An Internet diffusion framework. *Communications of the ACM, 41*(10): 21– 26.

Prime Minister's Strategy Unit. (2005). *Connecting the UK: The digital strategy.* London, United Kingdom.

Sclove, R. (2010). *Reinventing technology assessment: A 21st century model.* Washington, DC: Woodrow Wilson International Center for Scholars.

Selwyn, N., & Facer, K. (2007). *Beyond the digital divide: Rethinking digital inclusion for the 21st century.* Bristol, United Kingdom: Futurelab.

Settanni, C. (2007). *From conflict to collaboration: The "Wireless Minneapolis" community benefits agreement.* Presentation at the December 12, 2007, Community Development Forum, sponsored by the Federal Reserve Bank of Minneapolis & University of Minnesota's Center for Urban and Regional Affairs. Retrieved from http://digitalaccessproject.blogspot.com/2007/12/presentation-notes.html

Shaffer, G. (2007). Frame-Up: An analysis of arguments for and against municipal wireless initiatives. *Public Works Management Policy, 11*(3): 204–216.

Technology for All. (n.d.). *Texas Connects Coalition.* Retrieved from http://www.techforall.org/Programs/TXConnectsCoalitionTXC2/tabid/176/Default.aspx

U.S. General Accounting Office. (2003). *Information technology: A framework for assessing and improving enterprise architecture management (Version 1.1)*. Retrieved from http://www.gao.gov /new.items/ d03584g.pdf

Wireless Future Program, Stakeholder Strategy Committee. (2006). *First report: Wireless broadband and public needs: Building constituencies for spectrum policy change*. Chicago, IL: New America Foundation.

Wireless Internet Institute. (n.d.). *Digital inclusion forum*. Retrieved from http://www.digitalinclusionforum.com/

Wireless Minneapolis Digital Inclusion Task Force. (2006). *Final report*. Minneapolis, MN.

Reasons for Non-Adoption

Abraham, K. (2006). Nonresponse in the American time use survey: Who is missing from the data and How much does it matter? *Public Opinion Quarterly, 70*(5): 676–703.

Barzilai-Nahon, K. (2006). Gaps and bits: Conceptualizing measurements for digital divide/s. *The Information Society, 22*: 269–278. Retrieved from http://uwnews.washington.edu/ni/relatedcontent/2006/October/rc_parentID27342_thisID27365.pdf

Cushman, M., & Klecun, E. (2006a). How (can) nonusers engage with technology: Bringing in the digitally excluded. In E. Trauth, D. Howcroft, T. Butler, B. Fitzgerald, & J. DeGross (Eds.), *Social Inclusion: Societal and Organizational Implications for Information Systems* (347–364). Boston, MA: Springer.

_______. (2006b). *I've never tried it because I don't like it: Enabling technology choices*. Paper presented to the Information, Communication and Society Conference, York, United Kingdom, September 20–22, 2006.

Dailey, D., Byrne, A., Powell, A., Karaganis, J., & Chung, J. (2010). *Broadband adoption in low income communities*. Social Science Research Council. Retrieved from http://webarchive.ssrc.org/pdfs/ Broadband_ Adoption_v1.1.pdf

DiMaggio, P., Hargittai, E., Celeste, C., & Shafer, S. (2004). From unequal access to differentiated use: A literature review and agenda for research on digital inequality. In K. Neckermann (Ed.), *Social Inequality* (355–400). New York, NY: Russell Sage Foundation.

Dobransky, K., & Hargittai, E. (2006). The disability divide in Internet access and use. *Communication & Society 9,(*3): 313–334.

Ford, G. S., Koutsky, T., & Spiwak, L.J. (2008). *The broadband efficiency Index: What really drives broadband adoption across the OECD?* Phoenix Center Policy Paper No. 33. Washington, DC: Phoenix Center for Advanced Legal & Economic Public Policy Studies.

Hargittai, E., & Hinnant, A. (2008). Digital inequality: Differences in young adults' use of the Internet. *Communication Research, 35*(5): 602–621.

Hassani, S. N. (2006). Locating digital divides at home, work, and everywhere else. *Poetics, 34*(4–5): 250–272.

Hauge, J. A., & Prieger, J. E. (2010) Demand-side programs to stimulate adoption of broadband: What works? *Review of Network Economics, 9*(3).

Horrigan, J. B. (2007). *Closing the broadband divide.* Washington, DC: Pew Internet & American Life Project, Pew Research Center.

_______. (2009). *Stimulating broadband: If Obama builds it, will they log on?* Washington, DC: Pew Internet & American Life Project, Pew Research Center.

Jaeger, P. T. (2006). Telecommunications policy and individuals with disabilities: Issues of accessibility and social inclusion in the policy and research agenda. *Telecommunications Policy, 30*: 112–124.

Kvasny, L., & Keil, M. (2006). The challenges of redressing the digital divide: A tale of two U.S. cities. *Information Systems Journal, 16*(1): 23–53.

Livingstone, S., & Helsper, E. (2007). Gradations in digital inclusion: Children, young people, and the digital divide. *New Media & Society, 9*(4): 671–6.

Norris, P. (2001). *Digital divide: Civic engagement, information poverty, and the Internet worldwide.* Cambridge, United Kingdom: Cambridge University Press.

Notley, T., & Foth, M. (2008). Extending Australia's digital divide policy: An examination of the value of social inclusion and social capital policy frameworks. *Australian Social Policy, 7.*

Prieger, J., & Hu, W. M. (2008). The broadband digital divide and the nexus of race, competition, and quality. *Information Economics & Policy, 20*: 150– 167.

Rainie, L. (2010). *Broadband.* Keynote address at Colorado Broadband conference, Denver, November 15, 2010. Retrieved from http://www.pewinternet.org/~/media 0%20-%2011%2015%2010%20pdf%20-%20 Broadband%20- %20Colorado.pdf

Strover, S., Chapman, G., & Waters, J. (2004). The digital vicious cycle: Links between social disadvantage and digital exclusion in rural areas. *Telecommunications Policy, 28*(7/8): 465–485.

U.S. Department of Commerce, National Telecommunications and Information Administration. (2010a). *Digital nation: 21st century America's progress toward universal broadband Internet access.* Washington, DC.

_______. (2010b). *Exploring the digital nation: Home broadband Internet adoption in the United States.* Washington, DC

Verdegem, P., & Verhoest, P. (2009). Profiling the non-user: Rethinking policy initiatives stimulating ICT acceptance. *Telecommunications Policy, 33*: 642–652.

End Notes

[1] Definition based on Communities Connect Network, available at http://seattle.gov/tech/overview/What_is_Digital_Inclusion.pdf.

[2] The term "community" is understood here in its broadest sense. A community can be a geographic entity, but it can also reflect other characteristics, such as occupation, interest, or identity.

[3] "Anchor" institutions are community institutions that serve as a resource for broad segments of the population: schools, libraries, health care providers, public safety entities, community colleges and other institutions of higher education, and other support organizations.

In: Creating Digital Communities
Editor: Sloane Caruso
ISBN: 978-1-63117-631-9

Chapter 2

BROADBAND INTERNET ACCESS AND THE DIGITAL DIVIDE: FEDERAL ASSISTANCE PROGRAMS*

Lennard G. Kruger and Angele A. Gilroy

SUMMARY

The "digital divide" is a term that has been used to characterize a gap between "information haves and have-nots," or in other words, between those Americans who use or have access to telecommunications and information technologies and those who do not. One important subset of the digital divide debate concerns high-speed Internet access and advanced telecommunications services, also known as broadband. Broadband is provided by a series of technologies (e.g., cable, telephone wire, fiber, satellite, wireless) that give users the ability to send and receive data at volumes and speeds far greater than traditional "dial-up" Internet access over telephone lines.

Broadband technologies are currently being deployed primarily by the private sector throughout the United States. While the numbers of new broadband subscribers continue to grow, studies and data suggest that the rate of broadband deployment in urban/suburban and high income areas is outpacing deployment in rural and low-income areas. Some

* This is an edited, reformatted and augmented version of a Congressional Research Service publication, CRS Report for Congress RL30719, prepared for Members and Committees of Congress, from www.crs.gov, dated July 17, 2013.

policymakers, believing that disparities in broadband access across American society could have adverse economic and social consequences on those left behind, assert that the federal government should play a more active role to avoid a "digital divide" in broadband access.

With the conclusion of the grant and loan awards established by the American Recovery and Reinvestment Act of 2009 (P.L. 111-5), there remain two ongoing federal vehicles which direct federal money to fund broadband infrastructure: the broadband and telecommunications programs at the Rural Utilities Service (RUS) of the U.S. Department of Agriculture and the Universal Service Fund (USF) programs under the Federal Communications Commission (FCC). Although the USF's High Cost Program does not explicitly fund broadband infrastructure, subsidies are used, in many cases, to upgrade existing telephone networks so that they are capable of delivering high-speed services. Additionally, subsidies provided by USF's Schools and Libraries Program and Rural Health Care Program are used for a variety of telecommunications services, including broadband access. Currently the USF is undergoing a major transition to the Connect America Fund, which is targeted to the deployment, adoption, and utilization of both fixed and mobile broadband.

To the extent that the 113th Congress may consider various options for further encouraging broadband deployment and adoption, a key issue is how to strike a balance between providing federal assistance for unserved and underserved areas where the private sector may not be providing acceptable levels of broadband service, while at the same time minimizing any deleterious effects that government intervention in the marketplace may have on competition and private sector investment.

INTRODUCTION

The "digital divide" is a term used to describe a perceived gap between "information haves and have-nots," or in other words, between those Americans who use or have access to telecommunications and information technologies and those who do not.[1] Whether or not individuals or communities fall into the "information haves" category depends on a number of factors, ranging from the presence of computers in the home, to training and education, to the availability of affordable Internet access.

Broadband technologies are currently being deployed primarily by the private sector throughout the United States. While the numbers of new broadband subscribers continue to grow, studies and data suggest that the rate of broadband deployment in urban/suburban and high income areas is outpacing deployment in rural and low-income areas.

STATUS OF BROADBAND DEPLOYMENT IN THE UNITED STATES

Prior to the late 1990s, American homes accessed the Internet at maximum speeds of 56 kilobits per second by dialing up an Internet Service Provider (such as AOL) over the same copper telephone line used for traditional voice service. A relatively small number of businesses and institutions used broadband or high speed connections through the installation of special "dedicated lines" typically provided by their local telephone company. Starting in the late 1990s, cable television companies began offering cable modem broadband service to homes and businesses. This was accompanied by telephone companies beginning to offer DSL service (broadband over existing copper telephone wireline). Growth has been steep, rising from 2.8 million high speed lines reported as of December 1999, to 243 million connections as of June 30, 2012.[2] Of the 243 million high speed connections reported by the FCC, 197 million serve residential users.[3]

Table 1 depicts the relative deployment of different types of broadband technologies. A distinction is often made between "current generation" and "next generation" broadband (commonly referred to as next generation networks or NGN). "Current generation" typically refers to currently deployed cable, DSL, and many wireless systems, while "next generation" refers to dramatically faster download and upload speeds offered by fiber technologies and also by successive generations of cable, DSL, and wireless technologies.[4] In general, the greater the download and upload speeds offered by a broadband connection, the more sophisticated (and potentially valuable) the application that is enabled.

Based on the latest FCC broadband connection data, *Table 2* shows the percentages of households with broadband connections by state, both for download connections over 200 kbps and for connections of at least 3 Mbps (which approximates the FCC's National Broadband Availability target). According to the FCC, high speed connections over 200 kbps are reported in 67% of households nationwide, while connections of at least 3 Mbps (download) and 768 kbps (upload) are reported in 38% of households nationwide.

Table 1. Percentage of Broadband Technologies by Types of Connection

	Connections over 200 kbps in at least one direction	Residential connections over 200 kbps in at least one direction	Connections at least 3 Mbps downstream and 768 kbps upstream	Residential connections at least 3 Mbps downstream and 768 kbps upstream
cable modem	20.4%	23.9%	37.7%	45.3%
DSL	12.8%	14.1%	12.8%	14.7%
Mobile wireless	63.0%	58.2%	42.9%	32.5%
Fiber	2.6%	3.0%	6.0%	7.2%
All other	1.2%	0.8%	0.6%	0.4%

Source: FCC, Internet Access Services: Status as of June 30, 2012, pp. 23-26.

Table 2. Percentage of Households With Broadband Connections by State (as of June 30, 2012)

	Connections over 200 kbps	Connections at least 3 mbps downstream and 768 kbps upstream
Alabama	57%	30%
Alaska	71%	34%
Arizona	69%	54%
Arkansas	54%	25%
California	76%	44%
Colorado	76%	61%
Connecticut	79%	58%
Delaware	78%	69%
District of Columbia	72%	65%
Florida	74%	51%
Georgia	65%	42%
Hawaii	*	*
Idaho	61%	30%
Illinois	67%	44%
Indiana	64%	41%
Iowa	66%	25%
Kansas	67%	33%
Kentucky	62%	38%
Louisiana	60%	30%
Maine	75%	26%
Maryland	76%	70%
Massachusetts	81%	75%
Michigan	65%	46%

	Connections over 200 kbps	Connections at least 3 mbps downstream and 768 kbps upstream
Minnesota	69%	48%
Mississippi	50%	21%
Missouri	62%	29%
Montana	64%	38%
Nebraska	68%	45%
Nevada	66%	50%
New Hampshire	81%	65%
New Jersey	82%	76%
New Mexico	59%	37%
New York	76%	54%
North Carolina	69%	17%
North Dakota	70%	41%
Ohio	68%	25%
Oklahoma	58%	28%
Oregon	69%	55%
Pennsylvania	71%	58%
Rhode Island	76%	*
South Carolina	64%	29%
South Dakota	67%	47%
Tennessee	58%	39%
Texas	65%	33%
Utah	73%	56%
Vermont	77%	61%
Virginia	70%	60%
Washington	72%	61%
West Virginia	60%	49%
Wisconsin	67%	30%
Wyoming	66%	44%
National subscribership ratio	69%	45%

Source: FCC, Internet Access Services: Status as of June 30, 2012, pp. 34-35.

Notes: Asterisk (*) indicates data withheld by the FCC to maintain firm confidentiality. Subscribership ratio is the number of reported residential high speed lines (broadband connections) divided by the number of households in each state.

Meanwhile, the National Broadband Map, which is composed of state broadband data and compiled by NTIA, provides data on where broadband is and is not available. The latest update of these data indicate that 98.2% of the U.S. population have available advertised speeds of at least 3 Mbps (download) and 768 kbps (upload), 96.2% have download speeds 6 Mbps or greater, and 94.4% have download speeds 10 Mbps or greater.[5] The FCC's *Eighth Broadband Progress Report*, released on August 21, 2012, used National Broadband Map data to estimate that 19 million Americans living in 7 million households lack access to fixed broadband at speeds of 4 Mbps (download)/1 Mbps (upload) or greater.[6] *Table 3* shows a state-by-state breakdown of the percentage of population without access to fixed broadband at the FCC's benchmark speed of 4 Mbps/1Mbps.

Table 3. Americans Without Access to Fixed Broadband by State (access to speeds of at least 4 Mbps download/1 Mbps upload)

	% of population without access	% of population without access, nonrural areas	% of population without access, rural areas
United States	6.0%	1.8%	23.7%
Alabama	11.4%	1.6%	25.5%
Alaska	19.6%	4.4%	48.9%
Arizona	4.7%	1.2%	35.8%
Arkansas	13.6%	1.8%	28.8%
California	3.3%	1.6%	35.2%
Colorado	4.3%	1.0%	25.3%
Connecticut	0.7%	0.5%	2.6%
Delaware	3.1%	1.1%	13.0%
District of Columbia	0.0%	0.0%	N/A
Florida	3.1%	2.0%	14.3%
Georgia	3.4%	1.3%	9.9%
Hawaii	1.5%	0.1%	17.7%
Idaho	13.1%	1.3%	41.4%
Illinois	3.3%	0.4%	25.6%
Indiana	4.3%	1.3%	12.4%
Iowa	7.1%	0.7%	18.7%
Kansas	7.7%	1.0%	27.0%
Kentucky	10.5%	1.5%	23.0%
Louisiana	8.8%	1.3%	29.6%
Maine	4.7%	1.2%	7.0%
Maryland	3.2%	0.9%	19.2%
Massachusetts	1.0%	0.5%	6.4%
Michigan	6.3%	0.8%	22.4%
Minnesota	8.0%	0.8%	27.7%
Mississippi	12.1%	1.2%	22.8%
Missouri	7.5%	0.6%	24.2%
Montana	26.7%	4.0%	55.4%

	% of population without access	% of population without access, nonrural areas	% of population without access, rural areas
Nebraska	10.1%	1.9%	33.0%
Nevada	2.3%	0.6%	30.2%
New Hampshire	7.5%	2.5%	15.2%
New Jersey	0.7%	0.4%	5.6%
New Mexico	14.2%	4.8%	46.7%
New York	1.3%	0.0%	10.4%
North Carolina	6.4%	2.1%	15.0%
North Dakota	15.9%	2.5%	36.2%
Ohio	3.4%	0.5%	14.0%
Oklahoma	16.2%	2.9%	42.5%
Oregon	3.4%	0.2%	17.3%
Pennsylvania	1.7%	0.3%	6.8%
Rhode Island	0.2%	0.0%	2.3%
South Carolina	11.7%	4.9%	25.1%
South Dakota	21.1%	3.2%	44.6%
Tennessee	6.8%	0.9%	18.6%
Texas	5.9%	2.0%	27.6%
Utah	1.8%	0.3%	16.7%
Vermont	9.4%	0.2%	15.2%
Virginia	10.9%	2.2%	37.6%
Washington	3.2%	0.5%	17.4%
West Virginia	45.9%	31.4%	59.8%
Wisconsin	6.9%	0.1%	23.0%
Wyoming	13.2%	1.1%	35.4%
U.S. Territories	54.0%	41.5%	85.2%
American Samoa	78.6%	30.9%	92.0%
Northern Marianas	100.0%	100.0%	100.0%
Guam	54.3%	0.1%	76.1%
Puerto Rico	51.6%	40.3%	84.8%
U.S. Virgin Islands	100.0%	100.0%	100.0%

Source: FCC, Eighth Broadband Progress Report, Appendix C.

In contrast to broadband *availability*, which refers to whether or not broadband service is offered, broadband *adoption* refers to the extent to which American households actually subscribe to and use broadband. The U.S. Department of Commerce, based on October 2012 survey data, found that 72.4% of U.S. households have adopted broadband.[7] Similarly, the FCC's *Eighth Broadband Progress Report* found that 64% of American households with broadband available to them adopt broadband service offering speeds faster than 768 kbps/200 kbps, while 40% adopt speeds faster than the FCC benchmark of 4 Mbps/1Mbps. The FCC found that the "broadband adoption rates for American households are lower, on average, in the counties with the lowest median household income, in areas outside of urban areas, on Tribal lands, and in U.S. Territories."[8]

According to the Department of Commerce report *Exploring the Digital Nation: America's Emerging Online Experience* (based on July 2011 U.S. Census Bureau survey data), the three main reasons cited for not having broadband at home are that it is perceived as not needed, too expensive, and lack of a home computer.[9] The Department of Commerce report, the FCC's National Broadband Plan, and a survey conducted by the Pew Internet and American Life Project[10] also found disparities in broadband adoption among demographic groups. Populations continuing to lag behind in broadband adoption include people with low incomes, seniors, minorities, the less-educated, non-family households, and the non-employed.

BROADBAND IN RURAL AREAS[11]

While the number of new broadband subscribers continues to grow, the rate of broadband deployment in urban areas appears to be outpacing deployment in rural areas.

While there are many examples of rural communities with state of the art telecommunications facilities,[12] recent surveys and studies have indicated that, in general, rural areas tend to lag behind urban and suburban areas in broadband deployment. For example:

- According to the FCC's *Eighth Broadband Progress Report*, released in August 2012, of the 19 million Americans who live where fixed broadband is unavailable, 14.5 million live in rural areas.[13]
- According to June 2012 data from the National Broadband Map, 99.6% of the population in urban areas have access to available broadband download speeds of at least 6 Mbps, as opposed to 81.8% of the population in rural areas.[14]
- The Department of Commerce report, *Exploring the Digital Nation: America's Emerging Online Experience*, found that while the digital divide between urban and rural areas has lessened since 2007, it still

persists with 72% of urban households adopting broadband service in 2011, compared to 58% of rural households.[15]

The comparatively lower population density of rural areas is likely the major reason why broadband is less deployed than in more highly populated suburban and urban areas. Particularly for wireline broadband technologies—such as cable modem and DSL—the greater the geographical distances among customers, the larger the cost to serve those customers. Thus, there is often less incentive for companies to invest in broadband in rural areas than, for example, in an urban area where there is more demand (more customers with perhaps higher incomes) and less cost to wire the market area.[16]

Some policymakers believe that disparities in broadband access across American society could have adverse consequences on those left behind, and that advanced telecommunications applications critical for businesses and consumers to engage in e-commerce are increasingly dependent on high speed broadband connections to the Internet. Thus, some say, communities and individuals without access to broadband could be at risk to the extent that connectivity becomes a critical factor in determining future economic development and prosperity. A February 2006 study done by the Massachusetts Institute of Technology for the Economic Development Administration of the Department of Commerce marked the first attempt to quantitatively measure the impact of broadband on economic growth. The study found that "between 1998 and 2002, communities in which mass-market broadband was available by December 1999 experienced more rapid growth in employment, the number of businesses overall, and businesses in IT-intensive sectors, relative to comparable communities without broadband at that time."[17]

A June 2007 report from the Brookings Institution found that for every one percentage point increase in broadband penetration in a state, employment is projected to increase by 0.2% to 0.3% per year. For the entire U.S. private non-farm economy, the study projected an increase of about 300,000 jobs.[18]

Subsequently, a July 2009 study commissioned by the Internet Innovation Alliance found net consumer benefits of home broadband on the order of $32 billion per year, up from an estimated $20 billion in consumer benefits from home broadband in 2005.[19]

Some also argue that broadband is an important contributor to U.S. future economic strength with respect to the rest of the world. Data from the Organization for Economic Cooperation and Development (OECD) found the U.S. ranking 15th among OECD nations in broadband access per 100 inhabitants as of December 2011.[20] By contrast, in 2001 an OECD study found the U.S. ranking fourth in broadband subscribership per 100 inhabitants (after Korea, Sweden, and Canada).[21] While many argue that declining U.S. performance in international broadband rankings is a cause for concern,[22] others maintain that the OECD data undercount U.S. broadband deployment,[23] and that cross-country broadband deployment comparisons are not necessarily meaningful and are inherently problematic.[24] Finally, an issue related to international broadband rankings is the extent to which broadband speeds and prices differ between the United States and the rest of the world.[25]

Are Broadband Deployment Data Adequate?

Obtaining an accurate snapshot of the status of broadband deployment is problematic. Anecdotes abound of rural and low-income areas which do not have adequate Internet access, as well as those which are receiving access to high-speed, state-of-the-art connections. Rapidly evolving technologies, the constant flux of the telecommunications industry, the uncertainty of consumer wants and needs, and the sheer diversity and size of the nation's economy and geography make the status of broadband deployment very difficult to characterize. The FCC periodically collects broadband deployment data from the private sector via "FCC Form 477"—a standardized information gathering survey. Statistics derived from the Form 477 survey are published every six months. Additionally, data from Form 477 are used in the FCC's (to date) eight broadband deployment reports.

The FCC is working to refine the data used in future reports in order to provide an increasingly accurate portrayal. In its March 17, 2004, Notice of Inquiry for the *Fourth Report*, the FCC sought comments on specific proposals to improve the FCC Form 477 data gathering program.[26] On November 9, 2004, the FCC voted to expand its data collection program by requiring reports from all facilities based carriers regardless of size in order to better track rural and underserved markets, by requiring broadband providers to provide more information on the speed and nature of their service, and by establishing broadband-over-power line as a separate category in order to track its development and deployment. The FCC Form 477 data gathering program was extended for five years beyond its then March 2005 expiration date.[27]

On April 16, 2007, the FCC announced a Notice of Proposed Rulemaking which sought comment on a number of broadband data collection issues, including how to develop a more accurate picture of broadband deployment; gathering information on price, other factors determining consumer uptake of broadband, and international comparisons; how to improve data on wireless broadband; how to collect information on subscribership to voice over Internet Protocol service (VoIP); and whether to modify collection of speed tier information.[28]

On March 19, 2008, the FCC adopted an order that substantially expands its broadband data collection capability. Specifically, the order expands the number of broadband reporting speed tiers to capture more information about upload and download speeds offered in the marketplace, requires broadband providers to report numbers of broadband subscribers by census tract, and improves the accuracy of information collected on mobile wireless broadband deployment.

Additionally, in a Further Notice of Proposed Rulemaking, the FCC sought comment on broadband service pricing and availability.[29] The July 2009 data release (providing data as of June 30, 2008) was the final data set gathered under the old FCC Form 477. The February 2010 data report (December 31, 2008, data) reflected the new Form 477 data collection requirements. Meanwhile, during the 110th Congress, state initiatives to collect broadband deployment data in order to promote broadband in underserved areas were viewed as a possible model for governmental efforts to encourage broadband.

The Broadband Data Improvement Act was enacted by the 110th Congress and became P.L. 110-385 on October 10, 2008. The law requires the FCC to collect demographic information on unserved areas, data comparing broadband service with 75 communities in at least 25 nations abroad, and data on consumer use of broadband. The act also directs the Census Bureau to collect broadband data, the Government Accountability Office to study broadband data metrics and standards, and the Department of Commerce to provide grants supporting state broadband initiatives.

P.L. 111-5, the American Recovery and Reinvestment Act, provided NTIA with an appropriation of $350 million to implement the Broadband Data Improvement Act and to develop and maintain a national broadband inventory map. The National Broadband Map was first released on February 17, 2011 (http://www.broadbandmap.gov), and is updated every six months.[30]

Finally, the FCC's National Broadband Plan addressed the broadband data issue, recommending that the FCC and the U.S. Bureau of Labor Statistics (BLS) should collect more detailed and accurate data on actual availability, penetration, prices, churn, and bundles offered by broadband service providers to consumers and businesses, and should publish analyses of these data.

BROADBAND AND THE FEDERAL ROLE

The Telecommunications Act of 1996 (P.L. 104-104) addressed the issue of whether the federal government should intervene to prevent a "digital divide" in broadband access. Section 706 requires the FCC to determine whether "advanced telecommunications capability [i.e., broadband or high-speed access] is being deployed to all Americans in a reasonable and timely fashion."

Since 1999, the FCC has adopted and released eight reports pursuant to Section 706. The first five reports formally concluded that the deployment of advanced telecommunications capability to all Americans is reasonable and timely. Unlike the first five 706 reports, the sixth, seventh, and eighth reports concluded that broadband is not being deployed to all Americans in a reasonable and timely fashion. According to the *Eighth Broadband Progress Report*:

> Our analysis shows that the nation's broadband deployment gap remains significant and is particularly pronounced for Americans living in rural areas and on Tribal lands. We find that as of June 30, 2011, approximately 19 million Americans did not have access to fixed broadband. Significantly, approximately 76 percent of these Americans reside in rural areas. Our analysis further shows that Americans residing on Tribal lands disproportionately lack access to fixed broadband. And the available international broadband data, though not perfectly comparable to U.S. data, suggest that the availability and deployment of broadband in the United States may lag behind a number of other developed countries in certain respects, although we also compare favorably to some developed countries in other respects. Moreover, as many as 80 percent of E-rate recipients say that their broadband connections do not fully meet their needs, and 78 percent of recipients say that they need additional bandwidth. These data combined with our findings concerning availability above provide further indication that broadband is not yet being reasonably and timely deployed to all Americans.[31]

FCC Commissioners Robert McDowell and Ajit Pai issued dissenting statements, maintaining that there is insufficient justification for the 706 report conclusion that broadband is not being deployed in a reasonable and timely fashion. For example, the dissents argued that the report did not sufficiently account for the dramatic growth in the availability and deployment of mobile broadband, and that gaps in broadband adoption should not be used to determine whether or not broadband is being sufficiently deployed.[32]

The National Broadband Plan

As mandated by the ARRA, on March 16, 2010, the FCC publically released its report, *Connecting America: The National Broadband Plan.*[33] The National Broadband Plan (NBP) seeks to "create a high-performance America," which the FCC defines as "a more productive, creative, efficient America in which affordable broadband is available everywhere and everyone has the means and skills to use valuable broadband applications."[34] In order to achieve this mission, the NBP recommends that the country set six goals for 2020:

- Goal No. 1: At least 100 million U.S. homes should have affordable access to actual download speeds of at least 100 megabits per second and actual upload speeds of at least 50 megabits per second.
- Goal No. 2: The United States should lead the world in mobile innovation, with the fastest and most extensive wireless networks of any nation.
- Goal No. 3: Every American should have affordable access to robust broadband service, and the means and skills to subscribe if they so choose.
- Goal No. 4: Every American community should have affordable access to at least 1 gigabit per second broadband service to anchor institutions such as schools, hospitals, and government buildings.
- Goal No. 5: To ensure the safety of the American people, every first responder should have access to a nationwide, wireless, interoperable broadband public safety network.
- Goal No. 6: To ensure that America leads in the clean energy economy, every American should be able to use broadband to track and manage their real-time energy consumption.

The National Broadband Plan is categorized into three parts:

- *Part I (Innovation and Investment),* which "discusses recommendations to maximize innovation, investment and consumer welfare, primarily through competition. It then recommends more efficient allocation and management of assets government controls or influences."[35] The recommendations address a number of issues, including spectrum policy, improved broadband data collection, broadband performance standards and disclosure, special access rates, interconnection, privacy and cybersecurity, child online safety, poles and rightsof-way, research and experimentation (R&E) tax credits, and R&D funding.

- *Part II (Inclusion),* which "makes recommendations to promote inclusion—to ensure that all Americans have access to the opportunities broadband can provide."[36] Issues include reforming the Universal Service Fund, intercarrier compensation, federal assistance for broadband in Tribal lands, expanding existing broadband grant and loan programs at the Rural Utilities Service, enabling greater broadband connectivity in anchor institutions, and improved broadband adoption and utilization especially among disadvantaged and vulnerable populations.
- *Part III (National Purposes),* which "makes recommendations to maximize the use of broadband to address national priorities. This includes reforming laws, policies and incentives to maximize the benefits of broadband in areas where government plays a significant role."[37] National purposes include health care, education, energy and the environment, government performance, civic engagement, and public safety. Issues include telehealth and health IT, online learning and modernizing educational broadband infrastructure, digital literacy and job training, smart grid and smart buildings, federal support for broadband in small businesses, telework within the federal government, cybersecurity and protection of critical broadband infrastructure, copyright of public digital media, interoperable public safety communications, next generation 911 networks, and emergency alert systems.

The release of the National Broadband Plan is seen by many as a precursor towards the development of a national broadband policy—whether comprehensive or piecemeal—that will likely be shaped and developed by Congress, the FCC, and the Administration.[38] Congress will likely play a major role in implementing the National Broadband Plan, both by considering legislation to implement NBP recommendations, and by overseeing broadband activities conducted by the FCC and executive branch agencies.

CURRENT FEDERAL BROADBAND PROGRAMS

With the conclusion of grant and loan awards established by the American Recovery and Reinvestment Act of 2009 (P.L. 111-5),[39] there remain two ongoing federal vehicles which direct federal money to fund broadband infrastructure: the broadband and telecommunications programs at the Rural Utilities Service (RUS) of the U.S. Department of Agriculture and the Universal Service Fund (USF) programs under the Federal Communications Commission (FCC). Although the USF's High Cost Program does not explicitly fund broadband infrastructure, subsidies are used, in many cases, to upgrade existing telephone networks so that they are capable of delivering high-speed services. Additionally, subsidies provided by USF's Schools and Libraries Program and Rural Health Care Program are used for a variety of telecommunications services, including broadband access. Currently the USF is undergoing a major transition to the Connect America Fund, which is targeted to the deployment, adoption, and use of both fixed and mobile broadband.

Table 4 (at the end of this report) shows selected federal domestic assistance programs throughout the federal government that currently can be associated with broadband and telecommunications development. The table categorizes the programs in three ways: programs exclusively devoted to the deployment of broadband infrastructure; programs which have traditionally focused on deployment of telecommunications infrastructure generally (which typically can and does include broadband); and applications-specific programs which fund some aspect of broadband access or adoption as a means towards supporting a particular application, such as distance learning or telemedicine.

Rural Utilities Service Programs

RUS implements two programs specifically targeted at providing assistance for broadband infrastructure deployment in rural areas: the Rural Broadband Access Loan and Loan Guarantee Program and Community Connect Broadband Grants.[40] The 110th Congress reauthorized and reformed the Rural Broadband Access Loan and Loan Guarantee program as part of the 2008 farm bill (P.L. 110-234). The 112th Congress considered reauthorization of the program as part of the 2012 farm bill.[41]

RUS also has a rural telephone loan program (dating back to 1949, now called Telecommunications Infrastructure Loans) that has historically supported infrastructure for telephone voice service, but has now evolved into support for broadband-capable service provided by traditional telephone borrowers. Additionally, the Distance Learning and Telemedicine Grant Program supports broadband-based applications.[42]

The Universal Service Concept and the FCC[43]

Since its creation in 1934 the Federal Communications Commission (FCC) has been tasked with "mak[ing] available, so far as possible, to all the people of the United States ... a rapid, efficient, Nation-wide, and world-wide wire and radio communications service with adequate facilities at reasonable charges."[44] This mandate led to the development of what has come to be known as the universal service concept.

The universal service concept, as originally designed, called for the establishment of policies to ensure that telecommunications services are available to all Americans, including those in rural, insular and high cost areas, by ensuring that rates remain affordable. Over the years this concept fostered the development of various FCC policies and programs to meet this goal. The FCC offers universal service support through a number of direct mechanisms that target both providers of and subscribers to telecommunications and, more recently, broadband services.[45]

Universal Service and the Telecommunications Act of 1996

Passage of the Telecommunications Act of 1996 (P.L. 104-104) codified the long-standing commitment by U.S. policymakers to ensure universal service in the provision of telecommunications services.

The Schools and Libraries, and Rural Health Care Programs

Congress, through the 1996 act, not only codified, but also expanded the concept of universal service to include, among other principles, that elementary and secondary schools and classrooms, libraries, and rural health care providers have access to telecommunications services for specific purposes at discounted rates. (See §§254(b)(6) and 254(h)of the 1996 Telecommunications Act, 47 U.S.C. 254.)

1. The Schools and Libraries Program. Under universal service provisions contained in the 1996 act, elementary and secondary schools and classrooms and libraries are designated as beneficiaries of universal service discounts. Universal service principles detailed in Section 254(b)(6) state that

"Elementary and secondary schools and classrooms ... and libraries should have access to advanced telecommunications services." The act further requires in Section 254(h)(1)(B) that services within the definition of universal service be provided to elementary and secondary schools and libraries for education purposes at discounts, that is at "rates less than the amounts charged for similar services to other parties."

The FCC established the Schools and Libraries Division within the Universal Service Administrative Company (USAC) to administer the schools and libraries or "E (education)-rate" program to comply with these provisions. Under this program, eligible schools and libraries receive discounts ranging from 20% to 90% for telecommunications services depending on the poverty level of the school's (or school district's) population and its location in a high cost telecommunications area. The FCC established a funding ceiling, or cap, of $2.25 billion, adjusted for inflation prospectively beginning with funding year 2010. Three categories of services are eligible for discounts: internal connections (e.g., wiring, routers and servers); Internet access; and telecommunications and dedicated services, with the third category receiving funding priority. According to data released by program administrators, approximately $31 billion in funding has been committed over the first 14 years of the program with funding released to all states, the District of Columbia and all territories. Funding commitments for funding Year 2012 (July 1, 2012-June 30, 2013), the 15th year of the program, totaled $ 2.6 billion as of July 5, 2013.[46]

2. The Rural Health Care Programs. Section 254(h) of the 1996 act requires that public and nonprofit rural health care providers have access to telecommunications services necessary for the provision of health care services at rates comparable to those paid for similar services in urban areas. Subsection 254(h)(1) further specifies that "to the extent technically feasible and economically reasonable" health care providers should have access to advanced telecommunications and information services. The FCC established the Rural Health Care Division (RHCD) within the USAC to administer the universal support program to comply with these provisions. Under FCC established rules only public or non-profit health care providers are eligible to receive funding. Eligible health care providers, with the exception of those requesting only access to the Internet, must also be located in a rural area. The funding ceiling, or cap, for this support was established at $400 million annually. The funding level for Year One of the program (January 1998-June 30, 1999) was set at $100 million. Due to less than anticipated demand, the FCC established a $12 million funding level for the second year (July 1, 1999

to June 30, 2000) of the program but has since returned to a $400 million yearly cap. As of March 31, 2012, a total of $514.3 million has been committed since the program's inception in 1998. The primary use of the funding is to provide reduced rates for telecommunications and information services necessary for the provision of health care.[47] In addition, the FCC established, in 2007, the "Rural Health Care Pilot Program" to help public and non-profit health care providers build state and region-wide broadband networks dedicated to the provision of health care services. There are 50 projects in the program with $387.9 million in authorized funds. As of February 29, 2010, $232.6 million of the funds have been committed to the 50 FCC designated projects. The FCC in a December 12, 2012, order, created a new program, the Healthcare Connect Fund, which will expand health care provider access to broadband, particularly in rural areas, and replace the Rural Health Care Pilot Program with a permanent program. Non-rural participation is limited to consortia of which at least 50% must be located in a rural area. The upfront funding cap for the Healthcare Connect Fund is $150 million annually and participants will be required to contribute 35% of the costs. The FCC also established, as part of the Healthcare Connect Fund, a new pilot program, to expand broadband connections to skilled nursing facilities. Funding for this pilot program, which will begin in 2014, will be up to $50 million total over three years. The total funding cap for all of the above mentioned USF rural health care programs will remain at $400 million annually.[48]

Universal Service and Broadband

One of the policy debates surrounding universal service in the last decade was whether access to advanced telecommunications services (i.e., broadband) should be incorporated into universal service objectives. The term universal service, when applied to telecommunications, refers to the ability to make available a basket of telecommunications services to the public, across the nation, at a reasonable price.

As directed in the 1996 Telecommunications Act (§254[c]) a federal-state Joint Board was tasked with defining the services which should be included in the basket of services to be eligible for federal universal service support; in effect using and defining the term "universal service" for the first time. The Joint Board's recommendation, which was subsequently adopted by the FCC in May 1997, included the following in its universal service package: voice grade access to and some usage of the public switched network; single line service; dual tone signaling; access to directory assistance; emergency service such as 911; operator services; and access and interexchange (long distance) service.

Some policy makers expressed concern that the FCC-adopted definition is too limited and does not take into consideration the importance and growing acceptance of advanced services such as broadband and Internet access. They point to a number of provisions contained in the Universal Service section of the 1996 act to support their claim. Universal service principles contained in Section 254(b)(2) state that "Access to advanced telecommunications services should be provided to all regions of the Nation." The subsequent principle (b)(3) calls for consumers in all regions of the nation including "low-income" and those in "rural, insular, and high cost areas" to have access to telecommunications and information services including "advanced services" at a comparable level and a comparable rate charged for similar services in urban areas. Such provisions, they state, dictate that the FCC expand its universal service definition.

The 1996 act does take into consideration the changing nature of the telecommunications sector and allows for the universal service definition to be modified if future conditions warrant. Section 254(c)of the act states that "universal service is an evolving level of telecommunications services" and the FCC is tasked with "periodically" reevaluating this definition "taking into account advances in telecommunications and information technologies and services." Furthermore, the Joint Board is given specific authority to recommend "from time to time" to the FCC modification in the definition of the services to be included for federal universal service support. The Joint Board, on November 19, 2007, concluded such an inquiry and recommended that the FCC change the mix of services eligible for universal service support. The Joint Board recommended, among other things, that "the universal availability of broadband Internet services" be included in the nation's communications goals and hence be supported by federal universal service funds.[49]

This debate was put to rest when provisions contained in the American Recovery and Reinvestment Act of 2009 (ARRA) called for the FCC to develop, and submit to Congress, a national broadband plan to ensure that every American has "access to broadband capability."[50] The FCC in its national broadband plan, *Connecting America: the National Broadband Plan,* recommended that access to and adoption of broadband be a national goal. Furthermore the national broadband plan proposed that the Universal Service Fund be restructured to become a vehicle to help reach this goal. The FCC, in an October 2011 decision, adopted an Order that calls for the USF to be transformed, in stages, over a multi-year period, from a mechanism to support voice telephone service to one that supports the deployment, adoption, and utilization of both fixed and mobile broadband.

This transformation includes the phase out of the USF's legacy High Cost Program and the creation of a new fund, the Connect America Fund, to replace it.[51]

LEGISLATION IN THE 110TH CONGRESS

In the 110th Congress, legislation was enacted to provide financial assistance for broadband deployment. Of particular note is the reauthorization of the Rural Utilities Service (RUS) broadband loan program, which was enacted as part of the 2008 farm bill (P.L. 110-234). In addition to reauthorizing and reforming the RUS broadband loan program, P.L. 110-234 contains provisions establishing a National Center for Rural Telecommunications Assessment and requiring the FCC and RUS to formulate a comprehensive rural broadband strategy.

The Broadband Data Improvement Act (P.L. 110-385) was enacted by the 110th Congress and required the FCC to collect demographic information on unserved areas, data comparing broadband service with 75 communities in at least 25 nations abroad, and data on consumer use of broadband. The act also directed the Census Bureau to collect broadband data, the Government Accountability Office to study broadband data metrics and standards, and the Department of Commerce to provide grants supporting state broadband initiatives. Meanwhile, the America COMPETES Act (H.R. 2272) was enacted (P.L. 110-69) and contained a provision authorizing the National Science Foundation (NSF) to provide grants for basic research in advanced information and communications technologies. Areas of research included affordable broadband access, including wireless technologies. P.L. 110-69 also directs NSF to develop a plan that describes the current status of broadband access for scientific research purposes.

LEGISLATION IN THE 111TH CONGRESS

In the 111th Congress, legislation was enacted that sought to provide financial assistance for broadband deployment. Of particular note, provisions in the American Recovery and Reinvestment Act of 2009 (P.L. 111-5) provided grants and loans to support broadband access and adoption in unserved and underserved areas.

P.L. 111-5: The American Recovery and Reinvestment Act of 2009

On February 17, 2009, President Obama signed P.L. 111-5, the American Recovery and Reinvestment Act (ARRA). Broadband provisions of the ARRA provided a total of *$7.2 billion*, for broadband grants, loans, and loan/grant combinations.

The total consisted of $4.7 billion to NTIA/DOC for a newly established Broadband Technology Opportunities Program (grants) and $2.5 billion to the RUS/USDA Broadband Initiatives Program (grants, loans, and grant/loan combinations).[52]

Regarding the $2.5 billion to RUS/USDA broadband programs, the ARRA specified that at least 75% of the area to be served by a project receiving funds shall be in a rural area without sufficient access to high speed broadband service to facilitate economic development, as determined by the Secretary of Agriculture. Priority was given to projects that provide service to the most rural residents that do not have access to broadband services. Priority was also given to borrowers and former borrowers of rural telephone loans.

Of the $4.7 billion appropriated to NTIA:

- $4.35 billion was directed to a competitive broadband grant program, of which not less than $200 million shall be available for competitive grants for expanding public computer center capacity (including at community colleges and public libraries); not less than $250 million to encourage sustainable adoption of broadband service; and $10 million transferred to the Department of Commerce Office of Inspector General for audits and oversight; and
- $350 million was directed for funding the Broadband Data Improvement Act (P.L. 110-385) and for the purpose of developing and maintaining a broadband inventory map, which shall be made accessible to the public no later than two years after enactment. Funds deemed necessary and appropriate by the Secretary of Commerce may be transferred to the FCC for the purposes of developing a national broadband plan, which shall be completed one year after enactment.

Final BTOP and BIP program awards were announced by September 30, 2010.

Other Enacted Broadband Legislation in the 111th Congress

P.L. 111-8 (H.R. 1105). Omnibus Appropriations Act, 2009. Appropriates to RUS/USDA $15.619 million to support a loan level of $400.487 million for the Rural Broadband Access Loan and Loan Guarantee Program, and $13.406 million for the Community Connect Grant Program. To the FCC, designates not less than $3 million to establish and administer a State Broadband Data and Development matching grants program for state-level broadband demand aggregation activities and creation of geographic inventory maps of broadband service to identify gaps in service and provide a baseline assessment of statewide broadband deployment. Passed House February 25, 2009. Passed Senate March 10, 2009. Signed by President, March 12, 2009. P.L. 111-32 (H.R. 2346). Supplemental Appropriations Act, 2009. Provides not less than $3 million to the FCC to develop a national broadband plan pursuant to the American Recovery and Reinvestment Act of 2009. Introduced May 12, 2009; referred to Committee on Appropriations. Passed House May 14, 2009; passed Senate May 21, 2009. Signed by President, June 24, 2009. P.L. 111-80 (H.R. 2997). Agriculture, Rural Development, Food and Drug Administration, and Related Agencies Appropriations Act, 2010. For Rural Utilities Service, U.S. Department of Agriculture, provides $28.96 million to support a loan level of $400 million for the broadband loan program, and $17.97 million for broadband community connect grants. Introduced June 23, 2009; referred to Committee on Appropriations. Reported by Committee on Appropriations June 23, 2009. Passed House July 9, 2009. Passed Senate August 4, 2009. Conference Report (H.Rept. 111-279) printed September 30, 2009. Signed by President October 21, 2009.

LEGISLATION IN THE 112TH CONGRESS

The 112th Congress examined the efficacy of federal broadband assistance programs and how they may fit into the context of a national broadband policy.

P.L. 112-10 (H.R. 1473). Department of Defense and Full-Year Continuing Appropriations Act, 2011. Rescinds existing unobligated past-year funding for the Rural Broadband Access Loan and Loan Guarantee Program and the Community Connect Grants at the Rural Utilities Service. For FY2011, appropriates $22.3 million to the Rural Broadband Access Loan and Loan Guarantee Program for the cost of broadband loans, and $13.4 million to Community Connect Grants. Signed by President, April 15, 2011.

P.L. 112-55 (H.R. 2112). Consolidated and Further Continuing Appropriations Act, 2012. Provides FY2012 appropriations for Rural Utilities Service broadband loan program and broadband community connect grants: $6 million for the broadband loan program (subsidizing a loan level of $212 million) and $10.372 million for Community Connect grants. Introduced June 3, 2011; referred to Committee on Appropriations. Reported by Committee on Appropriations June 3, 2011 (H.Rept. 112-101). Passed House June 16, 2011. Reported by Senate Appropriations Committee September 7, 2011 (S.Rept. 112-73). Signed by President, November 18, 2011.

H.R. 1083 (Owens). Rural Broadband Initiative Act. Establishes an Office of Rural Broadband Initiatives in the Department of Agriculture which would administer the RUS broadband loan and grant programs, and would develop a comprehensive rural broadband strategy. Introduced March 15, 2011; referred to Committee on Agriculture and in addition to the Committee on Energy and Commerce.

H.R. 1343 (Bass). To return unused or reclaimed funds made available for broadband awards in the American Recovery and Reinvestment Act of 2009 to the Treasury of the United States. Introduced April 4, 2011; referred to Committee on Energy and Commerce and to Committee on Agriculture. Reported (amended) by the Committee on Energy and Commerce (H.Rept. 112-228) on September 29, 2011. Passed House October 5, 2011. Referred to Senate Committee on Commerce, Science and Transportation October 6, 2011.

H.R. 1695 (Eshoo). Broadband Conduit Deployment Act of 2011. Directs the Secretary of Transportation to require that broadband conduit be installed as part of certain highway construction projects. Introduced May 3, 2011; referred to Committee on Transportation and Infrastructure.

H.R. 2163 (Matsui). Broadband Affordability Act of 2011. Amends the Communications Act of 1934 to establish a Lifeline Assistance Program for universal broadband adoption. Introduced June 14, 2011; referred to Committee on Energy and Commerce.

H.R. 5973 (Kingston). Agriculture, Rural Development, Food and Drug Administration, and Related Agencies Appropriations Act, 2013. For Rural Utilities Service, U.S. Department of Agriculture, provides $2 million to support a loan level of $21 million for the broadband loan program, and $10 million for broadband community connect grants. Introduced June 20, 2012; referred to Committee on Appropriations. Reported by Committee on Appropriations June 20, 2012.

H.R. 6083 (Lucas). Federal Agriculture Reform and Risk Management Act of 2012. Reauthorizes rural broadband loan program at $25 million per year through FY2017. Introduced July 9, 2012; referred to Committee on Agriculture. Ordered to be reported by committee July 11, 2012.

S. 257 (Landrieu). Small Business Broadband and Emerging Information Technology Enhancement Act of 2011. Seeks to improve certain programs of the Small Business Administration to better assist small business customers in accessing broadband technology. Introduced February 2, 2011; referred to Committee on Small Business and Entrepreneurship.

S. 1659 (Ayotte). To return unused or reclaimed funds made available for broadband awards in the American Recovery and Reinvestment Act of 2009 to the Treasury of the United States. Introduced October 5, 2011; referred to Committee on Commerce, Science and Transportation.

S. 1939 (Klobuchar). Broadband Conduit Deployment Act of 2011. Directs the Secretary of Transportation to require that broadband conduit be installed as part of certain highway construction projects. Introduced December 1, 2011; referred to Committee on Environment and Public Works.

S. 2298 (Brown of Ohio). Connecting Rural America Act. Amends the Rural Electrification Act of 1936 to improve the program of access to broadband telecommunications services in rural areas. Introduced April 18, 2002; referred to Committee on Agriculture, Nutrition, and Forestry.

S. 2375 (Kohl). Agriculture, Rural Development, Food and Drug Administration, and Related Agencies Appropriations Act, 2013. For Rural Utilities Service, U.S. Department of Agriculture, provides $6 million to support a loan level of $63 million for the broadband loan program, and $10 million for broadband community connect grants. Introduced April 26, 2012; referred to Committee on Appropriations. Reported by Committee on Appropriations April 26, 2012.

S. 3240 (Stabenow). Agriculture Reform, Food, and Jobs Act of 2012. Authorizes broadband loan and grant program at $50 million per year through FY2017. Introduced May 24, 2012; referred to Committee on Agriculture, Nutrition and Forestry. Reported to Senate May 24, 2012. Passed Senate (amended) June 21, 2012.

S. 3439 (Snowe). Federal Wi-Net Act. Directs the Administrator of General Services to install Wi-Fi hotspots and wireless neutral host systems in all federal buildings in order to improve in-building wireless communications coverage and commercial network capacity by offloading wireless traffic onto wireline broadband networks. Introduced July 25, 2012; referred to Committee on Environment and Public Works.

LEGISLATION IN THE 113TH CONGRESS

The following is a listing of broadband legislation directly related to the issue of federal assistance for broadband deployment in unserved areas.[53]

P.L. 113-6 (H.R. 933). Consolidated and Further Continuing Appropriations Act, 2013. Funds the broadband loan program at $4 million (supporting a loan level of approximately $42 million) and the Community Connect grant program at $10.372 million. Signed by President March 26, 2013.

H.R. 2163 (Matsui). Broadband Adoption Act of 2013. Amends the Communications Act of 1934 to reform and modernize the Universal Service Fund Lifeline Assistance Program. Introduced April 26, 2013; referred to Committee on Energy and Commerce.

H.R. 1947 (Lucas). Federal Agriculture Reform and Risk Management Act of 2013. Section 6105 would reauthorize the broadband loan and loan guarantee program through FY2018 at the current level of $25 million per year. Introduced May 13, 2013; reported by Committees on Agriculture and Judiciary.

H.R. 1639 (Gibson). Amends the Rural Electrification Act of 1936 to authorize loan/grant combinations under RUS broadband program. Introduced April 18, 2013; referred to Committee on Agriculture and Committee on Energy and Commerce.

H.R. 2410 (Aderholt). Agriculture, Rural Development, Food and Drug Administration, and Related Agencies Appropriations Act, 2014. Provides $5.5 million to subsidize a loan level of $42.146 million for the broadband loan program, and $10.111 million for the Community Connect grant program. Introduced June 18, 2013; reported by House Committee on Appropriations (H.Rept. 113-116).

S. 954 (Stabenow). Agriculture Reform, Food, and Jobs Act of 2013. Includes the establishment of a new grant program in combination with the existing loan and loan guarantee program authorization, which is extended at $50 million per year through FY2018. Passed by Senate June 10, 2013.

Table 4. Selected Federal Domestic Assistance Programs Related to Broadband and Telecommunications Development

Program	Agency	Description	Funding Amount (est. FY2012 unless otherwise noted)	Web Links
		Broadband Infrastructure Deployment Programs		
Broadband Technology Opportunities Program (BTOP)	National Telecommunications and Information Administration, Dept. of Commerce	Provides competitive grants to public and private sector entities in order to provide broadband access in unserved and underserved areas; provide broadband support and services to strategic institutions; improve broadband access by public safety agencies; and stimulate broadband demand, economic growth, and job creation.	$4 billion (ARRA, P.L. 111-5) (2009)	http://www.ntia.doc.gov/ broadbandgrants/
Broadband Initiatives Program (BIP)	Rural Utilities Service, U.S. Dept. of Agriculture	Provides competitive grants, loans, and loan/grant combinations to public and private sector entities in order to provide broadband access in unserved and underserved rural areas.	$2.5 billion for the cost of loans, grants, and loan/grant combinations (ARRA, P.L. 111-5) (2009)	http://www.rurdev.usda.gov/utp_bip.html
Rural Broadband Access Loan and Loan Guarantee Program	Rural Utilities Service, U.S. Dept. of Agriculture	Provides loan and loan guarantees for facilities and equipment providing broadband service in rural communities	$169 million for cost of money loans	http://www.rurdev.usda.gov/utp_farmbill.html
Community Connect Broadband Grants	Rural Utilities Service, U.S. Dept. of Agriculture	Provides grants to applicants proposing to provide broadband service on a "community-oriented connectivity" basis to rural communities of under 20,000 inhabitants.	$10 million	http://www.rurdev.usda.gov/ utp_commconnect.html

Program	Agency	Description	Funding Amount (est. FY2012 unless otherwise noted)	Web Links
	Telecommunications Broadband Infrastructure Deployment Programs			
Telecommunications Infrastructure Loan Program	*Rural Utilities Service, U.S. Dept. of Agriculture*	Provides long-term direct and guaranteed loans to qualified organizations for the purpose of financing the improvement, expansion, construction, acquisition, and operation of telephone lines, facilities, or systems to furnish and improve telecommunications service in rural areas. All facilities financed must be capable of supporting broadband services.	$145 million for hardship loans; $250 million for cost of money loans; and $295 million for FFB Treasury loans	http://www.rurdev.usda.gov/utp_infrastructure.html
Distance Learning and Telemedicine Loans and Grants	*Rural Utilities Service, U.S. Dept. of Agriculture*	Provides seed money to rural community facilities (e.g., schools, libraries, hospitals) for advanced telecommunications systems that can provide health care and educational benefits to rural areas	$21 million	http://www.rurdev.usda.gov/ UTP_DLT.html
Universal Service High Cost Program[a]	*Federal Communications Commission*	Provides funding to eligible telecommunications carriers to help pay for telecommunications services in high-cost, rural, and insular areas so that prices charged to customers are reasonably comparable across all regions of the nation.	$4.5 billion (annually through 2017)	http://www.usac.org/hc/

Table 4. (Continued)

Program	*Agency*	Description	Funding Amount (est. FY2012 unless otherwise noted)	Web Links
Universal Service Schools and Libraries Program (i.e., E-rate)	*Federal Communications Commission*	Provides discounts for affordable telecommunications and Internet access services to ensure that schools and libraries have access to affordable telecommunications and information services.	$2.25 billion (annually adjusted for inflation)	http://www.universalservice.org/sl/
Universal Service Rural Health Care Pilot Program[b]	*Federal Communications Commission*	Provides funds to cover 85% of the cost of constructing statewide or regional broadband telehealth networks and of connecting those projects to dedicated nationwide broadband telehealth networks and the public Internet.	$418 million (annually)	http://www.usac.org/rhc-pilot-program/
Appalachian Area Development Program	*Appalachian Regional Commission*	Project grants to support self-sustaining economic development in the region's most distressed counties and areas. Includes funds for a Telecommunications Initiative involving projects that enable communities to capitalize on broadband access.	$56 million	http://www.arc.gov/index.do?nodeId= 21
States' Economic Development Assistance Program	*Delta Regional Authority*	Grants for self-sustaining economic development projects of eight states in Mississippi Delta region.	$11 million (2011)	http://grants.dra.gov/
Investments for Public Works and Economic	*Economic Development Administration,*	Provides funding for construction of infrastructure in areas that are not attractive to private investment; most funding is for	$112 million	http://www.eda.gov/

Program	*Agency*	Description	Funding Amount (est. FY2012 unless otherwise noted)	Web Links
Development Facilities	*Dept. of Commerce*	water and sewer infrastructure but some has been designated for telecommunications and broadband projects.		
Library Services and Technology Act Grants to States	*Institute of Museum and Library Services, National Foundation on the Arts and the Humanities*	Provides funds for a wide range of library services including installation of fiber and wireless networks that provide access to library resources and services.	$156 million	http://www.imls.gov/programs/programs.shtm
Native American Library Services	*Institute of Museum and Library Services, National Foundation on the Arts and the Humanities*	Grants to support library services including electronically linking libraries to networks.	$4 million	http://www.imls.gov/applicants/grants/nativeAmerican.shtm
Choice Neighborhood Implementation Grants	*Office of the Assistant Secretary for Public and Indian Housing and Office of Multifamily Housing Programs, Dept. of Housing and Urban Development*	Helps communities transform neighborhoods by revitalizing severely distressed public and/or assisted housing. Grantees may use funds to provide unit-based broadband Internet connectivity.	$110 million	http://www.hud.gov/cn/

Table 4. (Continued)

Program	*Agency*	Description	Funding Amount (est. FY2012 unless otherwise noted)	Web Links
Special Education—Technology and Media Services for Individuals with Disabilities	*Office of Special Education and Rehabilitative Services, Dept. of Education*	Supports development and application of technology and education media activities for disabled children and adults	$30 million	http://www.ed.gov/about/offices/list/osers/index.html?src=mr/
Telehealth Network Grants	*Health Resources and Services Administration, Department of Health and Human Services*	Grants to develop sustainable telehealth programs and networks in rural and frontier areas, and in medically unserved areas and populations.	$6 million	http://www.hrsa.gov/telehealth/
Telehealth Resource Center Grant Program	*Health Resources and Services Administration, Department of Health and Human Services*	Provides grants that support establishment and development of telehealth resource centers to assist health care providers in the development of telehealth services, including decisions regarding the purchase of advanced telecommunications services.	$4 million	http://www.hrsa.gov/telehealth/
Licensure Portability Grant Program	*Health Resources and Services Administration, Department of Health and Human*	Provides support for state professional licensing boards to develop and implement state policies that will reduce statutory and regulatory barriers to telemedicine.	$0.35 million	http://www.hrsa.gov/telehealth/

Program	*Agency*	Description	Funding Amount (est. FY2012 unless otherwise noted)	Web Links
	Services			
NLM Extramural Programs	*National Library of Medicine, National Institutes of Health,*	Provides funds to train professional personnel; strengthen library and information services; facilitate access to and delivery of	$62 million	http://www.nlm.nih.gov/ep/extramural.html
	Department of Health and Human Services	health science information; plan and develop advanced information networks; support certain kinds of biomedical publications; and conduct research in medical informatics and related sciences.		
National Environmental Information Exchange Network Grant Program	*Environmental Protection Agency*	Provides funding to states, territories, and federally recognized Indian Tribes to support the development of an Environmental Information Exchange Network, including broadband infrastructure.	$10 million	http://epa.gov/exchangenetwork/grants/

Source: Compiled by CRS from FY2013 budget documents, the Catalog of Federal Domestic Assistance, and grants.

[a] The High Cost program will be phased out, and replaced in stages by the Connect America Fund. The program provides funding to eligible service providers to support the provision of affordable voice and broadband services, both fixed and mobile.

[b] The Rural Health Care program will be replaced by the Healthcare Connect Fund.

Conclusion

To the extent that the 113th Congress may consider various options for encouraging broadband deployment and adoption, a key issue is how to strike a balance between providing federal assistance for unserved and underserved areas where the private sector may not be providing acceptable levels of broadband service, while at the same time minimizing any deleterious effects that government intervention in the marketplace may have on competition and private sector investment.

In addition to loans, loan guarantees, and grants for broadband infrastructure deployment, a wide array of policy instruments are available to policymakers, including universal service reform, tax incentives to encourage private sector deployment, broadband bonds, demand-side incentives (such as assistance to low income families for purchasing computers), regulatory and deregulatory measures, and spectrum policy to spur roll-out of wireless broadband services. In assessing federal incentives for broadband deployment, the 113th Congress may consider the appropriate mix of broadband deployment incentives to create jobs in the short and long term, the extent to which incentives should target next-generation broadband technologies, the extent to which "underserved" areas with existing broadband providers should receive federal assistance, and whether broadband stimulus projects are being efficiently managed and how they may fit into the context of overall goals for a national broadband policy.

End Notes

[1] The term "digital divide" can also refer to international disparities in access to communications and information technology. This report focuses on domestic issues only.

[2] FCC, Internet Access Services: Status as of June 30, 2012, released May 2013, p. 17. Available at http://transition.fcc.gov/Daily_Releases/Daily_Business/2013/db0520/DOC-321076A1.pdf.

[3] Ibid.

[4] Initially, and for many years following, the FCC defined broadband (or more specifically "high-speed lines") as over 200 kilobits per second (kbps) in at least one direction, which was roughly four times the speed of conventional dialup Internet access. In recent years, the 200 kbps threshold was considered too low, and on March 19, 2008, the FCC adopted a report and order (FCC 08-89) establishing new categories of broadband speed tiers for data collection purposes. Specifically, 200 kbps to 768 kbps is considered "first generation," 768 kbps to 1.5 Mbps is "basic broadband tier 1," and increasingly higher speed tiers are

broadband tiers 2 through 7 (tier seven is greater than or equal to 100 Mbps in any one direction). Tiers can change as technology advances.

[5] NTIA, U.S. Broadband Availability: June 2010 – June 2012, May 2013, p.5.available at http://www.ntia.doc.gov/files/ ntia/publications/usbb_avail_report_05102013.pdf.

[6] Federal Communications Commission, Eighth Broadband Progress Report, FCC 12-90, released August 21, 2012, p. 29, available at http://transition.fcc.gov/Daily_Releases/Daily_ Business/2012/db0827/FCC-12-90A1.pdf.

[7] U.S. Department of Commerce, National Telecommunications and Information Administration, Blog, "Household Broadband Adoption Climbs to 72.4 Percent," June 6, 2013, available at http://www.ntia.doc.gov/blog/2013/householdbroadband-adoption-climbs-724-percent.

[8] Eighth Broadband Progress Report, p. 54.

[9] U.S. Department of Commerce, National Telecommunications and Information Administration, Exploring the Digital Nation: America's Emerging Online Experience, June 2013, p. 26, available at http://www.ntia.doc.gov/files/ntia/ publications/exploring_the_digital_nation_-_americas_emerging_online_experience.pdf.

[10] Smith, Aaron, Pew Internet & American Life Project, Home Broadband 2010, August 11, 2010, available at http://www.pewinternet.org/~/media//Files/Reports/2010/ Home% 20broadband%202010.pdf.

[11] For more information on rural broadband and broadband programs at the Rural Utilities Service, see CRS Report RL33816, Broadband Loan and Grant Programs in the USDA's Rural Utilities Service, by Lennard G. Kruger.

[12] See for example: National Exchange Carrier Association (NECA), Trends 2006: Making Progress With Broadband, 2006, 26 p. Available at http://www.neca.org/media/trends_ brochure_website.pdf.

[13] Federal Communications Commission, Eighth Broadband Progress Report, FCC 12-90, released August 21, 2012, p. 5, available at http://transition.fcc.gov/Daily_Releases/ Daily_Business/2012/db0827/FCC-12-90A1.pdf.

[14] NTIA, National Broadband Map, Broadband Statistics Report: Broadband Availability in Urban vs. Rural Areas, January 2013, p. 7, available at http://www.broadbandmap.gov/ download/ Broadband%20Availability%20in%20Rural%20vs%20Urban%20Areas.pdf. Also see NTIA, U.S. Broadband Availability: June 2010–June 2012, May 2013, p. 10-11, available at http://www.ntia.doc.gov/files/ntia/publications/ usbb_avail_report_ 05102013. pdf.

[15] U.S. Department of Commerce, National Telecommunications and Information Administration, Exploring the Digital Nation: America's Emerging Online Experience, June 2013, p. 26, available at http://www.ntia.doc.gov/files/ ntia/publications/exploring_ the_digital_nation_-_americas_emerging_online_experience.pdf.

[16] The terrain of rural areas can also be a hindrance to broadband deployment because it is more expensive to deploy broadband technologies in a mountainous or heavily forested area. An additional added cost factor for remote areas can be the expense of "backhaul" (e.g., the "middle mile") which refers to the installation of a dedicated line which transmits a signal to and from an Internet backbone which is typically located in or near an urban area.

[17] Gillett, Sharon E., Massachusetts Institute of Technology, Measuring Broadband's Economic Impact, report prepared for the Economic Development Administration, U.S. Department of Commerce, February 28, 2006, p. 4.

[18] Crandall, Robert, William Lehr, and Robert Litan, The Effects of Broadband Deployment on Output and Employment: A Cross-sectional Analysis of U.S. Data, June 2007, 20 pp. Available at http://www3.brookings.edu/ views/papers/crandall/200706litan.pdf.

[19] Mark Dutz, Jonathan Orszag, and Robert Willig, The Substantial Consumer Benefits of Broadband Connectivity for U.S. Households, Internet Innovation Alliance, July 2009, p. 4, http://internetinnovation.org/files/special-reports/ CONSUMER_BENEFITS_OF_BROADBAND.pdf.

[20] OECD, OECD Broadband Portal. Available at http://www.oecd.org/sti/ict/broadband.

[21] OECD, Directorate for Science, Technology and Industry, The Development of Broadband Access in OECD Countries, October 29, 2001, 63 pp. For a comparison of government broadband policies, also see OECD, Directorate for Science, Technology and Industry, Broadband Infrastructure Deployment: The Role of Government Assistance, May 22, 2002, 42 pp.

[22] See Turner, Derek S., Free Press, Broadband Reality Check II: The Truth Behind America's Digital Divide, August 2006, pp. 8-11. Available at http://www.freepress.net/files/bbrc2-final.pdf; and Turner, Derek S., Free Press, 'Shooting the Messenger' Myth vs. Reality: U.S. Broadband Policy and International Broadband Rankings, July 2007, 25 pp., available at http://www.freepress.net/files/shooting_the_messenger.pdf.

[23] National Telecommunications and Information Administration, Fact Sheet: United States Maintains Information and Communication Technology (ICT) Leadership and Economic Strength, at http://www.ntia.doc.gov/ntiahome/press/ 2007/ICTleader_042407.html.

[24] See Wallsten, Scott, Progress and Freedom Foundation, Towards Effective U.S. Broadband Policies, May 2007, 19 pp. Available at http://www.pff.org/issues-pubs/pops/pop14.7 usbroadbandpolicy.pdf. Also see Ford, George, Phoenix Center, The Broadband Performance Index: What Really Drives Broadband Adoption Across the OECD?, Phoenix Center Policy Paper Number 33, May 2008, 27 pp.; available at http://www.phoenix-center.org/pcpp/PCPP33Final.pdf.

[25] See price and services and speed data on OECD Broadband Portal, available at http://www.oecd.org/sti/ict/broadband; see also Federal Communications Commission, Third Annual International Broadband Data Report, IB Docket No. 10-171, DA 12-1334, August 21, 2012, available at http://hraunfoss.fcc.gov/edocs_public/attachmatch/DA12-1334A1.pdf.

[26] Federal Communications Commission, Notice of Inquiry, "Concerning the Deployment of Advanced Telecommunications Capability to All Americans in a Reasonable and Timely Fashion, and possible Steps to Accelerate Such Deployment Pursuant to Section 706 of the Telecommunications Act of 1996," FCC 04-55, March 17, 2004, p. 6.

[27] FCC News Release, FCC Improves Data Collection to Monitor Nationwide Broadband Rollout, November 9, 2004. Available at http://hraunfoss.fcc.gov/edocs_public/ attachmatch/DOC-254115A1.pdf.

[28] Federal Communications Commission, Notice Proposed Rulemaking, "Development of Nationwide Broadband Data to Evaluate Reasonable and Timely Deployment of Advanced Services to All Americans, Improvement of Wireless Broadband Subscribership Data, and Development of Data on Interconnected Voice Over Internet Protocol (VoIP) Subscribership," WC Docket No. 07-38, FCC 07-17, released April 16, 2007, 56 pp.

[29] FCC, News Release, "FCC Expands, Improves Broadband Data Collection," March 19, 2008. Available at http://hraunfoss.fcc.gov/edocs_public/attachmatch/DOC-280909A1.pdf.

[30] For more information on the national broadband mapping program and the State Broadband Data and Development Program, see http://www.ntia.doc.gov/broadband grants/ broadbandmapping.html.

[31] Eighth Broadband Progress Report, p. 59-60.

[32] Ibid., p.171, 177.

[33] Available at http://www.broadband.gov/plan/For more information on the National Broadband Plan, see CRS Report R41324, The National Broadband Plan, by Lennard G. Kruger et al..

[34] Federal Communications Commission, Connecting America: The National Broadband Plan, March 17, 2010, p. 9.

[35] Ibid., p. 11.

[36] Ibid.

[37] Ibid.

[38] See for example, Office of Science and Technology Policy and National Economic Council, The White House, Four Years of Broadband Growth, June 2013, 26 pages, available at http://www.whitehouse.gov/sites/default/files/ broadband_report_final.pdf.

[39] See CRS Report R40436, Broadband Infrastructure Programs in the American Recovery and Reinvestment Act, by Lennard G. Kruger.

[40] For more information on these programs, see CRS Report RL33816, Broadband Loan and Grant Programs in the USDA's Rural Utilities Service, by Lennard G. Kruger.

[41] Ibid.

[42] See CRS Report R42524, Rural Broadband: The Roles of the Rural Utilities Service and the Universal Service Fund, by Angele A. Gilroy and Lennard G. Kruger.

[43] The section on universal service was prepared by Angele Gilroy, Specialist in Telecommunications, Resources, Science and Industry Division. For more information on universal service, see CRS Report RL33979, Universal Service Fund: Background and Options for Reform, by Angele A. Gilroy.

[44] Communications Act of 1934, As Amended, Title I §1 [47 U.S.C. 151].

[45] Many states participate in or have programs that mirror FCC universal service mechanisms to help promote universal service goals within their states.

[46] For additional information on this program, including funding commitments, see the E-rate website: http://www.universalservice.org/sl/.

[47] For additional information on this program, including funding commitments, see the RHCD website: http://www.universalservice.org/rhc/.

[48] For more details on the USF rural health care support mechanism and the newly established Healthcare Connect Fund see In the Matter of Rural Health Care Support Mechanism, WC Docket No. 02-60, Federal Communications Commission, adopted December 12, 2012. Available at http://hraunfoss.fcc.gov/edocs_public/attachmatch/FCC-12- 150A1.pdf.

[49] The Joint Board recommended that the definition of those services that qualify for universal service support be expanded and that the nation's communications goals include the universal availability of: mobility services (i.e., wireless voice); broadband Internet services; and voice services at affordable and comparable rates for all rural and non-rural areas. For a copy of this recommendation see http://hraunfoss.fcc.gov/edocs_public/ attachmatch/FCC-07J4A1.pdf.

[50] American Recovery and Reinvestment Act of 2009, P.L. 111-5, sec. 6001 (k)(2)(D).

[51] For a detailed discussion of this Order and USF transition see CRS Report R42524, Rural Broadband: The Roles of the Rural Utilities Service and the Universal Service Fund, by Angele A. Gilroy and Lennard G. Kruger.

[52] For information on existing broadband programs at RUS, see CRS Report RL33816, Broadband Loan and Grant Programs in the USDA's Rural Utilities Service, by Lennard G. Kruger.

[53] For information on public safety wireless broadband legislation, see CRS Report R41842, Funding Emergency Communications: Technology and Policy Considerations, by Linda K. Moore.

INDEX

F

G

H

I

J

K

L

M

N

O

P

Q

R

S

T

U

V

W

Y